SCHOOLCRAFT COLLEGE LIBRARY
W9-BAV-659

America Identified

America Identified

Biometric Technology and Society

Lisa S. Nelson

The MIT Press
Cambridge, Massachusetts
London, England

For information about special quantity discounts, please e-mail special_sales@mit-press.mit.edu

This book was set in Stone Sans and Stone Serif by Toppan Best-set Premedia Limited. Printed and bound in the United States of America.

Library of Congress Cataloging-in-Publication Data

Nelson, Lisa S. (Lisa Sue)
American identified: biometric technology and society / Lisa S. Nelson.
p. cm.
Includes bibliographical references and index.
ISBN 978-0-262-01477-9 (hardcover : alk. paper)
1. Technological innovations—Social aspects. 2. Biometric identification.
3. Privacy Right of. 4. Social interaction—Technological innovations. I. Title.
HM846.N45 2011
303.48′3—dc22

2010011956

10 9 8 7 6 5 4 3 2 1

Contents

Preface

When I set out to study societal perceptions of biometric technology, I did not expect that it would reveal so much about the beliefs, aspirations, and values of American society beyond attitudes about biometric technology. In the aftermath of the terrorist attacks of September 11, 2001, many policy advocates disparaged the public sentiment as reactionary and ill informed. My research, however, revealed a more informed and nuanced public concerned with business, government, and social institutions. I discovered a public that exercised thoughtful insight into the potential uses of biometric technology, as well as on the increasing role of information technology and the information that it generates and collects. This research teaches us that we must not forget the role of users and potential users in the creation and deployment of information technologies and that they are an important source of understanding for those entrusted with developing public policy.

My writing of this book was sustained by friends, family, mentors, and colleagues. I am first grateful to Edward Constant, who, through kindness, humor, intelligence, and insight, instilled in me a curiosity and passion for the sociology of technology. His thoughtful advice and criticism throughout the evolution of the manuscript was invaluable.

I reserve special thanks to my colleagues with whom I collaborated on the National Science Foundation grant that supported this research. Without the collegiality and contribution of these scholars, including George Trapp (West Virginia University, WVU), Bojan Cukic (WVU), Stephanie Schuckers (Clarkson), Michael Schuckers (St. Lawrence University), Anil Jain (Michigan State), and Larry Hornack (WVU), this research and book would not have been possible. In addition to the generous support of the National Science Foundation, I am indebted to the Center for Identification Technology Research at West Virginia University for its early support through which I was able to develop a framework for my research.

I am indebted to the assistance and diligence of Marguerite Avery at the MIT Press and her kind assistants, Erin Shoudy and Johna Picco, for their work on the manuscript. I owe a debt of gratitude to the anonymous reviewers whose thoughtful and sometimes provocative criticisms and insights challenged me to more fully develop this work.

I hold in special esteem Bernard Yack and Marion Smiley, who, as my professors as the University of Wisconsin–Madison, taught me much more than political theory. I also had the good fortune while in law school at the University of Wisconsin–Madison to come under the tutelage of Leonard Kaplan, who has long been an invaluable mentor to me. I hope many students find a professor like John Witte, who, as the chair of the Political Science Department at the University of Wisconsin–Madison, gave me the opportunity to pursue graduate studies.

There are also my classmates from graduate school from whom I learned in unexpected ways. My friend Julie White engaged in endless conversations with me about life, paternalism, liberalism, and all things political theory on long walks through the arboretum. Christopher Burke challenged me in more ways than one, for which I remain grateful.

A profound remembrance and thanks is reserved for my mother and father, Josephine and Ralph Nelson, who taught me that dreams are meant to be pursued. On a tobacco farm in southern Wisconsin, they schooled me in the value of hard work and tenacity, both of which were required for this book and for my life as a teacher, author, and mother.

I owe the greatest debt to my husband, Tom Lammert, who read many drafts of the manuscript and offered countless suggestions and insights, all the while encouraging me to do better. Without his intellectual and emotional contributions, this book would not be. I also thank my daughter, Emma Lammert, whose arrival in this world blessed me with new motivation, purpose, and perspective.

Introduction

There was of course no way of knowing whether you were being watched at any given moment. How often, or on what system, the Thought Police plugged in on any individual wire was guesswork. It was even conceivable that they watched everybody all the time. But at any rate they could plug in your wire whenever they wanted to. You had to live—did live, from habit that became instinct—in the assumption that every sound you made was overheard, and, except in darkness, every movement scrutinized.

—George Orwell, *1984* (1949)

Before 9/11, the idea that Americans would voluntarily agree to live their lives under the gaze of a network of biometric surveillance cameras, peering at them in government buildings, shopping malls, subways, and stadiums, would have seemed unthinkable, a dystopian fantasy of a society that had surrendered privacy and anonymity.

—Jeffrey Rosen, *The Naked Crowd: Reclaiming Security and Freedom in an Anxious Age* (2004)

The widespread use of biometric technology—child of science fiction, crime dramas, and political theorizations throughout much of the twentieth century—has become a reality. The number of biometric identifiers available for the purpose of authenticating identity is diverse and expanding, prompting a range of reactions, concerns, and regulatory challenges. Biometric technology uses a biometric pattern—physiological,[1] anatomical, or behavioral attributes—to link individuals to biometric patterns. These attributes, ever evolving, now include fingerprints, voice, iris, retina, hand, face, handwriting, and keystroke, among a list of emerging alternatives. In broadest terms, biometric technologies "are automated methods of verifying or recognizing the identity of a living person based on a physiological or behavioral characteristic" (Wayman et al. 2005, 16).

Some have viewed biometric technology optimistically; others see it cataclysmically. Critics of biometric technology predict the arrival of Orwell's *1984:* "For anyone who has read Orwell's Nineteen Eighty-Four, where telescreens keep track of people's lives, this new biometric technology will seem like fiction come to life. It's showing up everywhere" (2006, xx).

Those favoring biometric technology frame it as a technology of identification and verification with the potential to fend off a wide range of atrocities from terrorism to identity theft. As John Woodward (2001) posited, "Biometric-based systems will become increasingly important tools for identifying known and suspected terrorists. One tool to counter the threat of terrorism is the use of emerging biometric technologies" (1). And even with the passing of the immediacy of the terrorist attacks of September 11, 2001, the potential for biometric technology as a tool of identity management in the information age represents an expanding and perhaps even more viable market. For example, as market analyst Maxine Most (2008) argued,

> In this respect, the overarching concept of "identity protection" offers a much stronger foundation and compelling value proposition for developing biometrically enabled applications than simply "enhancing security." In addition, it moves the discussion of biometrics away from the 100% accuracy and reliability problem to how much of an improvement or cost savings this technology can provide over existing processes and systems. (26)

Controversy has accompanied the rising visibility of biometric identification systems, pitting civil rights advocates against those with plans to use the technology in a variety of public and private sector settings to accomplish policy mandates. Commercial, governmental, and forensic applications for biometric systems of identification appear to be limited only by the imagination of engineers, scientists, academicians, businesses, and security experts. Biometric identification systems have been deployed in the commercial environment for the purposes of enhancing data security or verification of identity for ATM transactions, physical access, medical records management, or network log-in. In the governmental sector, the more controversial applications of biometric identification systems are found in the proposal of a national identification card, welfare disbursement, border control, or passport management. Biometric identification is also part of a body of forensic systems used to identify missing children, determine parentage, and more generally investigate crimes.

Broader discussions about the challenge of regulating the use of biometric systems arise when these systems are used to regulate behavior by surveillance or as a form of identity assurance. Inherent in this discussion are questions regarding the nature of the governing norms, values, and expectations of society; the efficacy of law and policy to manage technology; the societal influences of biometric technology; and the ethical considerations that should shape the use of identification technologies. Biometric technology certainly serves a purpose in the growing need for identity assurance where personal information is the currency of social interaction, but how do we reconcile its use to ensure that fundamental rights and liberties of individuals are not compromised?

The answers to this question are diverse. Information technology and the information that it generates has increasingly become part of our daily lives, shaping our practices, discourses, and institutions in fundamental ways. Consumers, professionals, and organizations may use personal information to a variety of ends, and in perhaps limitless settings, raising novel questions of how to ensure institutional ethics and accountability. The escalating reliance on personal information not only challenges long-standing demarcations between public and private institution in terms of responsibilities, obligations, and limits, but also calls for a reconsideration of how to ensure the protection of long-standing values and normative ethics. The answer cannot be distilled down to merely following the law. It is possible to rely only on the existing legal guarantees that are embodied in constitutional guarantees, statutory guidelines, or contractual protections. But these protections are wanting in many respects, especially as the demarcations between the public and private sector blur in the acquisition and processing of personal information. Should we, in the alternative, listen to the advocates for the development of policy solutions? While the perspective of the advocates offers insight, often it is driven by a political agenda that distorts the policy suggestions, erring on either the side of prohibiting the use of biometric technology altogether in certain contexts or allowing the issue of privacy and civil liberties protection to be waylaid by the pursuit of market share. Worse yet, signs of a disconnect are growing between privacy advocates, who warn that privacy is in peril, and a new generation of individuals who view personal information as a form of currency in social networking, communication, and a means of empowerment. Given these points of debate, how can societal perceptions of biometric identification systems inform our thinking? My intention in this book is to use societal perceptions of biometric technology to map the public policy considerations that are important and consequential to the integration of the technology into our lives.

Technology as a Device

Perceptions of technology can serve as a means to understand the ideology of the users and potential users in the process of integrating and incorporating technology into our day-to-day lives. This study seeks to understand the values and norms—in other words, political ideology—that affect societal acceptance of biometric technology. A starting point for technological or policy innovation is to understand the individuals who are subject to its purview. In this research, the societal perceptions of biometric technology reveal that the paramount ideological imperatives of privacy, anonymity, trust, and confidence in institutions and the legitimacy of paternalistic intervention are necessary considerations in the effort to build a policy paradigm for the integration of biometric technology into the daily experiences of Americans.

Toward a Methodology of Understanding Users

That individuals matter to the development and deployment of technology is an idea grounded in the literature of sociology of technology. In this approach, technology and society are dynamic and interactive, each affecting the other in the construction of meaning given a technology. For example, the social construction of technology (SCOT) approach, made popular by Pinch and Bijker(1984), demonstrated that social groups can transform the meaning of technology, shaping and influencing its development and deployment, especially in the early stages of its evolution (Bijker and Pinch 1987). This approach to the study of technology acknowledges that technology does not arise isolated from the methods, interests, materials, and institutions that influence its constitution; rather, technology is an artifact that arises from the complex interaction of these social groups and interests. The importance of identities, institutions, discourses, and the politics of representation in understanding the interaction between society and technology is central to the SCOT approach: "It acknowledges that lived `reality' is made up of complex linkages among the cognitive, the material, the normative and the social" (Jasanoff 2004, 274). From this recognition that technology can be affected by and affects social groups came the realization that users of technology also matter and influence the uses of technology. As technology enters the worlds of the users, there is room for consumers to "modify, domesticate, design, reconfigure, and resist technologies" (Oudshoorn and Pinch 2005, 1). Without users, technology ceases to become a part of society, as Sismondo (2008) writes: "For scientific knowledge and technological artifacts to be successful, they must be made to fit their environments or their environments must be made to fit them. . . . Part of the work of successful technoscience, then, is the construction not only of facts and artifacts but also of the societies that accept, use, and validate them" (17).

The inquiry into the interface between technology and users has yielded insights into the dynamic flexibility of technology in its developmental stages. Alkrich (1992), for example, pays particular attention to the awareness on the part of the technologists to anticipate users and in doing so to build a script or a scenario that depicts competencies, actions, and responsibilities to users. Others have argued that while technologists may be able to anticipate some of the interests, skills, motives, and behaviors of users, there is also interpretive flexibility of the users. This active perspective of users sees technological innovation and change as a fluid process that shapes social identities and social life (Oudshoorn and Pinch 2003). Sharing this perspective of users, Fisher (1992), in his work on the telephone, for instance, observed that users integrate technology in ways not always consistent with the a priori speculation of its developers but also inconsistent with critics or proponents.

The possibility of widespread access to the telephone evoked a host of hopes and apprehensions (Fischer 1992). From either perspective, the telephone, it was argued, would forever alter the fabric of American life. The telephone industry, for its part, promoted the telephone as an instrument of sociability. A 1921 advertisement described it as a tool for promoting family and community life: "It's a weekly affair now, those fond intimate talks. Distance rolls away and for a few minutes every Thursday night the familiar voices tell the little family gossip that both are so so eager to hear" (Fischer 1992, 76). Although the industry said that "telephone calls enriched social ties, offering 'gaiety, solace, and security,' even making of

America a nation of neighbors," others were concerned that American community would forever be changed for the worse (Fischer 1992) and spoke about fears about the loss of privacy, the altered state of the American psyche, and the potential loss of community. According to Fischer, how the telephone might change the nature of personal relations was one of the primary concerns about its social implications. Some critics argued that the telephone "brought people into close contact but obliged them to live at wider distances and created a palpable emptiness across which voices seemed uniquely disembodied and remote" (Kern 1983, 216). This claimed characteristic of the telephone was not a benefit to community, these critics charged, but instead "another blow of modernity against *Gemeinschaft,* the close community" (Fischer 1992, 25). Nevertheless, users integrated this technology into their private and public worlds, not entirely altering their worlds but also not leaving them entirely unchanged. Despite early concerns preceding the widespread adoption of the telephone, the integration of the telephone into the lives of its users by the 1920s was described by Fischer as "nonchalant" and not nearly as dramatic as first anticipated:

> Most people saw telephoning as accelerating social life, which is another way of saying that telephoning broke isolation and augmented social contacts. A minority felt that telephones served this function too well. These people complained about too much gossip, about unwanted calls, or, as did some family patriarchs, about wives and children chatting too much. Most probably sensed that the telephone bell, besides disrupting their activities (as visitors might), could also bring bad news or bothersome requests. Yet only a few seemed to live in a heightened state of alertness, ears cocked for the telephone's ring—no more, perhaps, than sat anxiously alert for a knock on the door. (247)

For Fischer (1992) the user is important to the development of a social-constructivist approach that accounts for both "human agency and intentionality among end users" (17). The importance of human agency is central to understanding how users matter. As Oudshoorn and Pinch (2005) discuss, users have played a significant role in technology:

> There is no correct use for a technology. "What is an alarm clock for?" we might ask. "To wake us up in the morning," we might answer. But just begin to list all the uses to which an alarm clock can be put and you see the problem. An alarm clock can be worn as a political statement by a rapper; it can be used to make a sound on a Pink Floyd recording; it can be used to evoke laughter, as Mr. Bean does in one of his comic sketches as he tries to drown his alarm clock in his bedside water pitcher; it can be used to trigger a bomb; and yes, it can be used to wake us up. (1)

As the example illustrates, users can adopt technologies in ways unanticipated by their inventors and developers or resist them altogether. When the Rural Electrification Administration (REA) began to secure the rights of way for its power lines, for example, landowners did not accept this innovation willingly. Rural resisters resorted to the use of shotguns and severed utility poles to thwart the agency's efforts. In response to the resistance, the REA wooed the rural community with demonstrations, kitchen parties, and farm equipment tours to show the promise of electrical appliances (Kline 2005). Other methodological explorations of the user include the role of patient advocacy communities on the research trajectories of breast cancer research and contraceptive development (Van Kammen, 2005). Research questions associated with the role of users include understanding the definition of users, the constitution of groups of users, and the association of political movements and social movements with users (Oudshoorn and Pinch 2003).

Although some research has explored the effects of the political ideology of inventors on the development of technology, surprisingly little attention has been paid to what perceptions of what technology can tell us about the politics or political ideology of users and the referents they rely on to evaluate the political, legal, and social acceptability of a technology. For example, the work of W. Bernard Carlson on the motivation of Alexander Graham Bell and Gardiner Hubbard in the invention and development of the telephone does not stop at economic interest. Carlson (2001) writes, "Most histories of the telephone treat its invention and development in purely economic terms: the telephone was invented because Alexander Graham Bell, Gardiner Hubbard, and Bell's other backers wanted to make money and it was adopted by businesses because it increased efficiency" (27). Instead of pure economic motivation, Carlson argues that Hubbard saw the telephone as nothing less than a technology that could potentially counter the rise of big business and temper a newly emerging financial elite who threatened the survival of a democratic America. For Hubbard, the telephone offered information and technology to a rising middle class: "Critics were especially concerned that Western Union had access to both market information and private business messages, and that the firm could use this information to manipulate markets in its favor and ruin individual businessmen" (Carlson 2001, 31).

Carlson, who examined the telephone as an artifact, reveals much about Hubbard's ideological predilections. Hubbard hoped that the advent of the telephone would not only undermine the monopolistic control of the telegraph by Western Union but would also empower the middle class with access to information, carving out a social space for the middle class and undermining the control of Western Union, which dominated communications prior to the invention of the telephone. Carlson argues that in the context of democracy, the telephone for Hubbard signaled progress for a middle class struggling against newly emerging monopolies and a financial elite. Used in this manner, a technological artifact like the telephone can also tell us something about the political ideology and political beliefs of the inventors of the technology.

Hecht (2001) offers another insight into politics, using the term *technopolitics* to refer to the "practice of designing or using technology to constitute, embody, or enact political goals" (285). In this sense, regimes use technology to accomplish political goals and construct a political identity. According to Hecht, the development of a nuclear program, for example, can be conceived of as part of a political strategy and the construction of national identity. His analysis illustrates that culture and politics are grounded in the materiality of technology and that the practices of technopolitics "reveal that the political power of culture cannot be separated from its material power" (287). Technology was used as tool for politics, and in the process, it revealed politics.

These limited inquiries into how technology may have been shaped by the politics, ideologies, and goals of inventors and governments lend insight into the research agenda of this book. The use of technology to inquire into the politics or political ideology of users or potential users can then be used to inform public policy. Social acceptance of a technology is more than a matter of a configuration of the technology for the users or placement of the technology within an existing legal paradigm. Instead, the question is how the politics or political ideology of users or potential users figure into the conditions of success or failure of a technology and how we can use the technology to understand political ideology.

Defining Ideology

The use of political ideology in this research is meant to evoke a discussion about the ideology, or preunderstandings in the sense that Heidegger might have implied, that is, serving as a means for individuals to evaluate their world. Without these preunderstandings "we would not be able to even identify any issue or understanding, let alone pass judgment upon it" (Eagleton 1991). *Political ideology,* as used here, Eagleton continues, does not refer to a "schematic, inflexible way of seeing the world, as against some more modest, piecemeal, pragmatic wisdom" (4). Understanding the political ideology of users or potential users is not accomplished at the level of the "high church," resting on an abstract notion of ideology, or at the level of critics or advocates, but rather at the level of the person on the street. This does not mean that norms and values are a function of discourse but that the discourse of ideas is part and parcel of the political ideology of users, and descriptions of ideology at the level of the individual are illustrative of a political reality. This is to say that an understanding of the effects of political ideology on the social acceptance of biometric technology must take the description of the concerns and hopes of users and potential users at face value. This idea of political ideology does not demean the ideology of users and potential users with the description of "false consciousness" by saying that those in the know (here, critics and advocates) better understand the social implications of biometric technology than do those in society. The notion of ideology at work here is not a theory of knowledge separated from the beliefs and values of individuals in society. But it does not follow that these perspectives are simply fleeting political perspectives. *Political ideology* in the sense used here is a process of signification. If a user views the deployment of biometric technology at a public event as unacceptable, what is signified or revealed by this rejection? Are the values of privacy and anonymity at play? Is the lack of trust and confidence in institutions significant? Is the presence of paternalistic intervention too substantial? Discussions of biometric technology and its potential uses reflect more than the mere words of users and potential users; they capture a political reality of long-standing beliefs and values that are part of political, cultural, and legal traditions.

Inquiring into the societal perceptions of biometric technology can tell us a great deal about the political values and beliefs of users and potential users and about the existing social, political, and intellectual context into which biometric technology is introduced. And it is not only perceptions of biometric technology that are revealed. These perceptions connect to enduring legal, social, and political considerations and can be used to outline the values, norms, and expectations that facilitate or present an obstacle to the integration and acceptance of systems of identification. The landscape of political ideology is expansive when it comes to biometric technology, in part because of the wide variety of biometric identifiers and their applications in diverse institutional settings. This diversity means that in some situations, the type of biometric identifier might not be acceptable because of its potential revelations about disease, as is the case with retinal scans. In other situations, social acceptance might not be given because the use of the identification technology is thought of as "overkill." It is helpful to understand the basics of biometric technology at the outset to better come to terms with all of the issues that can arise for users, potential users, or institutional actors.

The overarching task of a biometric system of identification is to verify an individual's identity or discover an individual's identity. In verification mode, biometric identification can be used to "test one of only two possible hypotheses: (1) that the submitted samples are from an individual known to the system; or (2) that the submitted samples are from an individual not known to the system" (Wayman et al. 2005, 5). Applications to test the first hypothesis, positive identification systems that verify a positive claim of enrollment, and applications testing the latter, negative identification, systems, verify a claim of no enrollment. The use of biometric identifiers to verify an individual's identity is accomplished by a process of comparison. The system confirms an individual's identity by comparing it to that individual's existing biometric template of a limited universe of individuals (Jain 2003).

There are many examples of positive identification systems. For instance, a number of states have used biometric identification to combat fraud in welfare payments. Beginning in 1994, Los Angeles County in California teamed with Printrak (an automatic fingerprint identification system) to provide the Automated Fingerprint Image Reporting and Match (AFIRM) system. A new welfare applicant applying for benefits must give a fingerprint of the forefinger of each hand. This person's personal information is stored in a database and is linked to a data storage subsystem that holds the fingerprint image data. In its initial assessment of the program, an audit report of the U.S. Department of Health and Human Services' Office of Inspector General entitled "Review of the Ongoing Los Angeles County Fingerprinting Demonstration Project" described the program a success in combating fraud. Similarly, in Ohio, several schools use fingerprints to identify students who receive free or reduced-price school lunches. School official describe "iMeal" as a way to remove stigma from students who receive the lunches, eliminating the need to hand over a ticket for each meal.

The greater reliability, accuracy, efficiency, and social acceptance of this type of one-to-one matching system makes it a more viable alternative than tokens and has given rise to a growing presence of biometric identification systems in commercial settings. Users of such technology tend to view it as convenient and a good means of protecting against identity theft, and the institutional actors are driven by considerations of preventing fraud.

Large-scale identification applications, which include border control, voter ID cards, driver's licenses, and national identification cards, pose different challenges. Highlighted in the aftermath of September 11, these one-to-many matching applications designed to screen individuals are problematic because of their accuracy and reliability and the issue of social acceptability: Even if a biometric identification system has a 99 percent success rate when applied to 200,000 people, there would still be a 1 percent failure rate, resulting in 200 false alarms.

A number of factors can undermine this use of biometric identification system in a one-to-many matching scenario— lack of enrollment of the user, minimal control over subjects, and less-than-ideal imaging conditions, for example—and each of these can result in misidentification. This type of biometric identification system also raises concerns related to civil rights because it often does not involve voluntary consent, as in the case of mandatory screening. The process may be covert, used in public settings, or viewed as an unnecessary exercise of surveillance with far too many risks for misidentification.

Social apprehensions are triggered by the use of biometric identification systems that relate to the rise of information technology generally. The increasing collation of personal information, including biometric identifiers, gives rise to a generalized concern of a loss of privacy and control associated by the creation of a digital dossier. "Digital technology enables the preservation of the minutia of our everyday comings and goings, of our likes and dislikes, of who we are and what we own. It is ever more possible to create an electronic collage that covers much of a person's life—a life captured in records, a digital person composed in the collective computer networks of the world" (Solove 2004, 1). What Solove describes as the "digital dossier" of personal information and the potential misuse serves as the backdrop for the increasing concerns regarding biometric technology. Even more troubling, innovation in information technology and the information it generates is creating an ever expanding universe of uses for personal information in public and private settings.

The concerns and aspirations associated with biometric technology tell us not only how to approach the development and deployment of biometric technology, but also something about the role of technology in the lives of individuals and how to create a system of regulation. Within this research are discussions about the value system, or political ideology, that is implicated when biometric technologies, and information technologies generally are used—an amalgam of something old, something borrowed, and something new. First, although political ideologies are not static, they are enduring. If we explore the implications of privacy when biometric technology is used, what can we learn about perceptions of privacy? Is the nature of privacy changing, and if so, how and why? Second, if the use of biometric technology is transformative of political ideologies, what can we learn about the regulation of new and existing applications of the technology by listening to users and potential users? Third, if biometric technology is used as a regulatory tool in the private sector or by government, is it a step toward the dystopian reality Orwell depicted, or it simply a new reality? Fourth, how can we use biometric technology to understand existing regulation and devise future regulation in the form of policy or technological innovation?

The Chapters: An Overview

Chapter 1 begins with a consideration of systems of identification from a historical perspective. Since the inception of the modern state, the issue of identifying its citizenry for a wide variety of purposes has been an imperative. For this reason, the use of biometric technology in the public and private sectors is in some ways not new because it is part of a lengthy historical discussion about the role of systems of identification.

Systems of identification have long been used to manage bureaucratic regimes, fend off external threats, and deal with classes of individuals thought to be dangerous to internal social stability. The concerns associated with the establishment of a system of identification are familiar too. Foremost among these are the loss of liberty, whether in the form of governmental control or the misuse of information, and the potential for loss of privacy and anonymity; all of which must be offset by the legitimacy of identification. External threats of war, identification of criminal elements in society, distribution of social welfare, and internal threats to social stability such as immigration often lead to public support of systems of identification. But when the prospect of long-term reliance on identification for the day-to-day bureaucratic workings of the government is the goal, the support of the citizenry wanes. Historically the factors of reliability and efficiency also have been important to establishing social acceptability. A system of identification that is prone to errors and resulting injustices does not retain public support. Nevertheless, agreement about the best forms of identification is far from a settled question even now. From the use of names, to fingerprinting, to anthropometry, systems of identification have been plagued by problems compromising of accuracy, reliability, and efficiency. A historical perspective on the use of these systems can lend insight into their future.

Chapter 2 considers the implications of a world altered by the information age and September 11. Touted as a panacea for identifying terrorists, biometric technology was introduced into an existing debate about the balance of liberty and security as the war on terror unfolded. September 11 set into motion an interest in large-scale deployment of biometric technology and at the same time fanned growing concerns regarding privacy and civil liberties. These highly politicized debates were not only about biometric technology. The use of this technology in the war on terror served to highlight an already growing controversy about the changing nature of society, culture, and politics in the information age in which the currency of personal information was increasingly being used in both the public and private sectors. This information revolution prompted discussions about the implications for traditional notions of individual liberty and the need to rethink policies. The availability of information technology and the increased reliance on it in society was already well underway prior to September 11 when biometric technology took center stage. To the debate, September 11 added questions about the proper use of surveillance and information technologies in the hands of government and the need to strengthen security without infringing on liberty.

Chapter 3 considers the relationship between privacy and biometric technology. Privacy concerns are central to the deployment of biometric technology and have figured prominently in public policy debates over the deployment of the technology. However, exactly how biometric technology affects and is affected by perceptions of privacy is complicated, largely because privacy is more than the language of legal doctrine. There are, of course, some protections that can be extended through statutes such as the Health Insurance Portability Act (HIPAA) or the Gramm-Leach-Bliley Act in the sectors of, respectively, medical or financial privacy or even through a broad interpretation of the Privacy Act of 1974 as it applies to governmental agencies. And the U.S. Constitution offers significant protections connected to the idea of privacy in terms of the realm of intimate decision making and the Fourth Amendment. These legal forms of privacy are not the end of understanding privacy in relationship to biometric technology, however.

As a normative value, privacy is a powerful factor in understanding societal acceptance of biometric technology. Where legal doctrine might not indicate a loss of privacy, societal perceptions might signify otherwise. And societal perceptions might support an acceptance of biometric technology even when privacy advocates argue against it. The normative values of privacy in consideration of biometric technology, accurate or not, must be understood as something complementary to, and perhaps influential in, the legal doctrine of privacy. Biometric technology does not implicate the end of privacy. Indeed, it may play a role in protecting privacy by serving as a tool of identity assurance in protecting the exchange of the currency of personal information. Increasingly, privacy in the information age must be considered in concert with the increasing reliance on the currency of personal information to participate in social interactions. Individual interest in privacy of personal information in this sense is not simply girding ourselves from the authority of the state or commercial entities. Privacy as a social practice is not built with isolation but instead requires confirmation and validation of an individual's moral agency in the realm of interactions. Even if biometric technology can serve a role in protecting the currency of personal information, considerations of the moral autonomy of the individual should still predominate. The protection of privacy in a changing realm of social interactions in the public and private sectors facilitates the exchange of personal information while protecting the realm of individual liberty.

Chapter 4 considers biometric technology as part of a larger debate about the impending loss of anonymity and decisional autonomy in a public sphere altered by tools of surveillance that are deployed in the public and private sectors. Many view the presence of surveillance technologies as undermining a sense of anonymity, or privacy in public, with the power of observation, evoking comparisons to Orwell's 1984 or Bentham's Panopticon, the architectural design of a prison that led prisoners to believe they were under constant surveillance. Conceptually the idea of anonymity and decisional autonomy have served important purposes in preserving some degree of private space for political opposition and the expression of ideas in the public sphere. Both the right to anonymity and the right of freedom of association are protected under the First Amendment, serving to define the public space of the associational life that underpins civil society. Surveillance technologies are considered a potential threat to these two pillars of liberty, potentially chilling the expression of ideas and constraining associational activity with observation. Historically there have been other times when changes have been viewed as detrimental to the dynamics of the public sphere. For instance, Jürgen Habermas argued that the structural change of the public sphere in the modern era prompted by the rise of state capitalism, the culture industries, and the increasingly powerful positions of economic corporations and big business in public life would undermine the vitality of public life. "Publicity loses its critical function in favor of a staged display," noted Habermas (1989). "Even arguments are transmuted into symbols to which again one cannot respond by arguing but only by identifying with them" (206). He lamented the demise of the "bourgeois public sphere" that provided social spaces for organization against arbitrary and domineering forms of social and public power. Yet these concerns did not materialize with quite the force he predicted. Although economic and governmental organizations took their place in the public sphere, citizens were not relegated to the role of mere consumers of goods, services, and political administration.

The rise of the information age and the events of September 11 have prompted a range of similar concerns as critics lament the demise of the public sphere in the face of surveillance technologies and a burgeoning use of our personal information by the public and private sectors. Indeed, we may be on the cusp of a new public sphere and, by the necessity of logic, a changed nature of associational life. The attendance of surveillance technologies on the day-to-day life of individuals may change the nature of the public sphere, but its destruction is unlikely. In fact, civil society is not completely devoid of state power, and the presence of state power is necessary to the preservation of civil society. Some state control, in the form of surveillance, laws, or police power, must exist in order to create the conditions of security that allow the freedom necessary for associational life to flourish.

The debate over the rightful place of surveillance technologies like biometric identification systems, and their incumbent effect on individual liberty in civil society, must consider the implications of a world and state power altered by these powerful technological tools. Participation in the public sphere requires the integration of personal information in our interactions within civil society but also with state authority. The idea of social freedom requires the presence of state authority, ensuring security in ways that may include technological observation to combat crime, fraud, and terrorism. Yet the increasing attendance of surveillance technologies in our lives does not translate neatly into a loss of individual liberty because the loci of power are not unilateral. The introduction of these complexities is not meant to minimize the social, cultural, and political implications of surveillance technologies. Instead, the accommodation of information technologies in the more traditional notions of state and civil society involves a reconsideration of the nature of state authority and the medium of civil society; both components are dependent on and constructed with technological means. The quandary is not to banish the effects of surveillance from our consideration of state authority or civil society but to discuss thoughtfully the implications of surveillance for state authority and civil society.

Chapter 5 considers the role of trust and confidence in the societal assessment of biometric technology. Since before the debates between the Federalists and the Anti-Federalists, there has been continuing consideration of building and maintaining trust between American institutions and society. Trust plays an important role in civic engagement, societal function, and organizational settings. More recent interest in trust has been in accounting for the effects of its loss and considering the alternatives to it. "The stunning collapse of Enron, WorldCom, and Arthur Andersen, and the Catholic church scandals forced a nationwide search for answers to fundamental questions about trust and trustworthiness" (Kramer and Cook 2004, 2). This search for trust was intensified by September 11, marking it as a potential turning point in America's perceived antipathy and distrust. Robert Putnam, widely known for his work on trust, hypothesized that perhaps

> September's tragedy opened a historic window of opportunity for civic renewal. Americans are more united, readier for shared sacrifice, and more public-spirited than in recent memory. Indeed, most American adults are experiencing their broadest-ever sense of "we." The World Trade Center disaster generated compelling cross-class images and cross-ethnic solidarity, linking the fates of Latino dishwashers, Irish firemen, and Jewish financiers. (Sander and Putnam 2002)

September 11 did more than heighten the need for greater trust and confidence in government institutions to prevent another terrorist attack; it also opened the door for claims of distrust because an expanded governmental role was built on the justification of the war on terror. The flow of personal information and the threats of identity theft and loss of privacy prompted public debates about an appropriate regime of regulation that would boost consumer confidence. But the potential for the misuse of personal information, already on the public agenda, took on greater importance after September 11 as revelations of secret surveillance programs began to undermine public confidence. AT&T's participation with the National Security Agency on a domestic surveillance program fanned criticism of both private and governmental institutions and their use of personal information and surveillance technologies. Appropriate use of technology became an important consideration of the function of trust in institutions: "If trust in institutions depends on the type of technology in use, trust in technology is also a function of level of trust in institutions that use the technology to begin with" (Zuriek and Hindle 2004, 118). Understanding perceptions of biometric technology is revealing of trust and confidence not only in the technology but also in the institutions making use of the technology. In this way, societal perceptions of biometric identification systems can tell us much about the role of trust and confidence in institutions where the presence of biometric technology is a part of the interface with the public and an increasingly important means of ensuring identity assurance.

Chapter 6 considers the influence of paternalism in the regulation and use of biometric technology. Although it is clear that trust and confidence play a part in societal acceptance of the use of personal information for institutional purposes, other philosophical influences are at play in societal perceptions of biometric technology. Information technology is a regulatory tool, serving to tighten the potential control a state may exercise over society. For this reason, a reconsideration of an enduring philosophical debate about legitimate paternalistic intervention is pertinent here. The expectation that government should exercise its power, though not entirely at the expense of individual liberty, in order to distribute benefits and to prevent harm is a well-established principle of democratic governance, made even more important in an era of a regulator regime using new technologies. The principle of paternalism, a topic of debate among foundational democratic philosophers such as John Stuart Mill, became more prominent during the New Deal, when the government's regulatory intervention to protect the citizenry gained greater legal currency.

The regulating influences of technologies, including biometric technology, designed to contribute to the security of information, government, transactions, and society, fall squarely in a tradition of public policy that in certain circumstances calls for institutions to make decisions on behalf of the citizenry. As individuals exercise choice over how to negotiate the privacy of their personal information, cyberpaternalists are already aligned in favor of paternalistic measures to regulate this free market approach. Discussions of broadly construed privacy protections like those exemplified by the European Union Data Directive or the guidelines of the Organization for Economic Cooperation and Development designed to influence institutions to act responsibly were at the forefront of political debates even prior to September 11.[2] In the aftermath of the terrorist attacks, concerns about the misuse of personal information grew, and calls for paternalistic regulation rose. The events of September 11 in fact changed the direction of the debate. In the aftermath of the terrorist attacks, the need for paternalistic intervention to combat the threat of terrorism in the forms of surveillance, information gathering, and identity authentication was juxtaposed against the rhetoric of regulation and protection of personal information on behalf of individuals. The importance of paternalism to societal perceptions of biometric technology cuts two ways.

Paternalism, in the acceptance and rejection of biometric technology in certain circumstances, is an important explanatory factor in understanding the range of societal perceptions. It sheds light on the circumstances within which individuals are willing to accept the presence of regulating technologies for the achievement of a benefit or the prevention of harm even when there is a potential risk to individual liberty. Paternalism may also explain the response of individuals with regard to the use of their personal information for purposes that they cannot see or control. Ironically, paternalism also explains why the use of biometric technology is not blindly accepted by the public. The legitimacy of paternalistic intervention is weighed against the risk being avoided. If the exercise of paternalistic intervention is viewed as threat to a liberty interest without legitimate justification, the public will likely not accept its exercise. Also important are the reliability and accuracy of the personal identifiers used, potential risks, the procedural guarantees put in place should mistakes occur, and the overarching question of whether paternalistic intervention is necessary to temper the security measures undertaken in the name of protecting against harm.

The Research Study

The goal of the research was to understand the complexity of societal reactions provoked by the various types and usages of biometric technology, making use of focus groups and a national survey. The sampling and statistical research on which this work rests used focus groups and a national survey. Focus groups enable the uncovering of a large range of views on a topic and allow a number of participants to share their many ideas in a short amount of time. Also, because the group dynamic often leads to conversational exchanges that show and challenge how people arrive at their views, the focus groups can be even more revealing of participants' thinking than what might emerge in a individual interviews.

The core objective of the qualitative focus group study was to better understand the views of biometric users and nonusers on a variety of issues related to how private information is protected and used by different institutions and to understand how biometric technology can potentially safeguard that private information and provide security. The focus group discussions examined four topics:

1. Privacy of personal information
2. Institutions and their handling of private information
3. Biometric technology as a way to protect private information
4. Situations where the use of biometric technology is acceptable

Some of the key questions included:

- Who collects personal information, and what do they collect?
- How well is private information safeguarded by different institutions, and how do they use it?
- Which types of personal information are safest to use for personal identification by an institution to safeguard privacy?
- What kinds of biometric technologies do the participants use? (This question was addressed to user groups only.)
- How safe and reliable are various types of biometric technology relative to other forms of identification?
- How acceptable is the use of biometrics for identification in various situations to protect information and provide security?

Recruitment of Participants

Focus group participants were recruited by posting public advertisements and contacting users of biometric technology identified through a database of previous study participants. Potential participants were screened to determine whether they were qualified for the study and were disqualified for the following reasons:

- They had participated in a focus group or any type of research study within the previous three-month period.
- They worked in building or information security.
- They did not speak English (the focus groups were conducted in English only).

Individuals who reported regularly using biometric technologies (referred to as users) who qualified for the study were scheduled for focus groups with other biometric users. Individuals who did not report regularly using biometrics (referred to as nonusers) were scheduled for focus groups with other nonusers.

During the screening process, we asked participants what biometric technologies they used and how often they used specific technologies.

Most users used biometric technology frequently (daily or a couple of times per week). A small minority had used biometric technology only a couple of times per year. Handprint and fingerprint were the most commonly used biometrics. Participants reported working in a variety of industries, including the military, private business (self-employed), technology, child care, health care, administrative, sales, education, and the service sector. None of the participants reported being unemployed, and only four reported that they were retired.

Thirty-two nonusers for four separate nonuser focus groups were recruited. We recruited fifty-five users to ensure the appropriate level of attendance at five separate user focus groups. Due to these recruiting efforts, each of the focus groups contained seven or eight participants. One focus group on a college campus morphed into a large-group discussion containing twenty-seven students as students willingly joined an open discussion.

Study Environment

Nine focus groups were conducted (four with nonusers and five with users), with sessions conducted at the American Institute for Research in Concord, Massachusetts; a conference room at the University of New Hampshire in Durham, New Hampshire; and a hotel conference room in Pineville, North Carolina.

All of the focus group rooms were outfitted with video and audio recording devices that captured the verbal interactions between participants and the moderator. Prior to the participants' arrival, the conference room was arranged to ensure appropriate lighting and air temperature and that participants had an adequate amount of table space. In addition, we provided pens for participants to use to complete brief surveys. During the sessions a moderator and note taker were present. The moderator facilitated the discussion, while the note taker prepared the room and documented the participants' comments.

The focus groups were a mix of survey completion and discussion. The surveys evaluated the four specific topics mentioned in the focus group objectives above (privacy of personal information, institutions and their handling of private information, biometric technology as a way to protect private information, and situations where the use of biometric technology is acceptable). In each session, participants completed each of the four surveys prior to the in-depth discussion on the topic to which the handout corresponded; this procedure helped ensure that participants voiced their own opinions, allowed a sense of individual perspective, and guaranteed that the discussion could be supplemented with quantitative data.

To begin the focus group session, the moderator asked the members which public or private institutions collected their personal information. After the group had volunteered some institutions, the moderator asked what types of personal information was collected. and then distributed the first of the four surveys used in each focus group. This one asked participants to rank their confidence in the role institutions play in keeping their personal information safe. (The surveys used a Likert scale from 1 to 5 in order to effectively measure the attitudes of both nonusers and users of biometric technology.) After all of the participants had completed the survey, the moderator asked each participant how he or she had ranked the various institutions listed. The moderator then guided participants through a series of questions on the handling of their private information relating to these following topics:

- Information sharing
- Hacking
- Identity theft
- Access to private information from unauthorized individuals
- Institutions most vulnerable to identity theft

The second survey listed a set of personal identifiers (e.g., social security number, date of birth) that institutions use to protect personal, private information. Participants rated how safe they thought each identifier was in protecting personal information from access by unauthorized persons. Next, the moderator asked participants to discuss their responses in greater detail and explain why certain types of information seemed safer than others to them. To conclude the discussion, the moderator asked participants how safe they believed fingerprints and voiceprints were compared to their social security number, a question used to make the transition of the conversation from the use of private information and the handling of that information by institutions to one about biometric technology.

Prior to distributing the next survey, the moderator asked participants of the nonuser groups what they had "heard" or "knew" about biometric technology. In the user groups, the moderator asked participants what biometric technology they used, how safe they believed it was, and their general feelings about their biometric use as it related to safety and security. The moderator then distributed the third survey, which asked participants to rate how effective a set of biometric identifiers was in protecting information from unauthorized access. When the participants had completed the survey, the moderator asked them to discuss why they rated certain items the way they did and whether they felt the use of biometric technology kept their private information safer than the traditional identifiers illustrated in the second survey.

Next, the moderator asked participants to rate how acceptable the use of biometric technology was in different situations. This fourth survey outlined a number of situations that related to both protection of private information (e.g., withdrawing money from a bank, making a credit purchase) and general security (e.g., boarding a plane, accessing government buildings). The ensuing discussion then focused on situations where biometric technology could be used. The moderator asked participants to consider how acceptable they believed the use of biometric technology was in certain situations.

To conclude the discussion, the moderator asked participants if they would like to see more frequent and expanded use of biometrics in the future.

The National Survey

The purpose of the national telephone survey, carried out in 2007, was to gather public opinion data on the issues of privacy or personal information and biometric technology from a representative sample of adults in the United States. Major topics covered in the survey included:

- The importance of protecting personal information
- Perceived threats to the privacy of information
- Comfort level with privacy protection offered by various institutions (e.g., employers, banks, the government)
- Comfort level with privacy protection offered by various security measures to identify individuals (e.g., passwords, mother's maiden name)
- Attitudes toward the government's role in protecting privacy
- Understanding of and attitudes toward biometric technology

This final sample was composed of 1,000 adults over the age of eighteen years, living in the United States.

Sample Characteristics

Respondents ranged in age from 18 to 93 years, with a mean age of 51.63 years (sd = 16.85 years). Slightly more females participated in the survey than males: 57.6 percent of the sample was female and 42.3 percent male. (Table I.1 shows other sample characteristics.)

When asked to indicate whether they had ever been a victim of identify theft, only 16.6 percent of the sample reported being a victim. When asked whether they had ever used biometric technology, 18.8 percent of respondents reported they had used biometrics. Most biometric users reported being infrequent users of the technology: 53.2 percent reported using it

Table I.1
Demographic characteristics of the telephone survey sample (*N* = 1,000)

Characteristic	Frequency (%)
Race/ethnicity	
White, including Hispanic	83.3
Black or African American	7.3
Hispanic, nonwhite	3.2
Asian or Pacific Islander	2.0
Other	1.4
Missing information or refused to answer	2.8
Highest educational attainment	
Eighth grade or less	1.3
Some high school, did not graduate	5.9
High school diploma or GED	22.9
Some college or two-year degree	29.6
Four-year college degree	20.2
More than four-year college degree	19.2
Missing information or refused to answer	0.9
Political affiliation	
Democrat	28.6
Republican	26.9
Independent	21.6
None	10.6
Other	1.2
Missing information or refused to answer	8.8

Table I.2
Frequency of biometric use (N = 188)

Characteristic	Frequency (%)
A couple of times per year	53.2
A couple of times per month	14.9
Weekly	6.4
A couple of times per week	3.7
Daily	15.4
Don't know	6.4

only a couple of times per year, and 15.4 percent of biometric users reported using it daily. Table I.2 summarizes the frequency with which biometric users reported using biometric technology.

Study Environment and Protection of Human Subjects

All survey data were collected by telephone. While introducing the survey to potential respondents, the caller explained that participation in the study was voluntary and that respondents had the right to refuse to answer any questions.

Conclusion

The research revealed a complex mosaic of societal impressions of biometric technologies that cannot be separated from the values, norms, and beliefs that give rise to them. Coming to terms with the societal acceptance of biometric technologies turns on a comprehension of the context of the present and the influence of the past. Systems of identification, including biometric identification systems, should be understood as part of a continuum of competing objectives rather than a revolutionary event.

1 Modern Identification Systems

He had a rich abundance of idle time, but it never hung heavy on his hands, for he interested himself in every new thing that was born into the universe of ideas, and studied it, and experimented upon it at his house. One of his pet fads was palmistry. To another one he gave no name, neither would he explain to anybody what its purpose was, but merely said it was an amusement. In fact, he had found that his fads added to his reputation as a pudd'nhead; there, he was growing chary of being too communicative about them. The fad without a name was one which dealt with people's finger marks. He carried in his coat pocket a shallow box with grooves in it, and in the grooves strips of glass five inches long and three inches wide. Along the lower edge of each strip was pasted a slip of white paper. He asked people to pass their hands through their hair (thus collecting upon them a thin coating of the natural oil) and then making a thumb-mark on a glass strip, following it with the mark of the ball of each finger in succession. Under this row of faint grease prints he would write a record on the strip of white paper—thus: JOHN SMITH, right hand—and add the day of the month and the year, then take Smith's left hand on another glass strip, and add name and date and the words "left hand." The strips were now returned to the grooved box, and took their place among what Wilson called his "records."

—Mark Twain, *Pudd'nhead Wilson* (1894)

Whether for the purpose of solving a murder as in Mark Twain's *Pudd'nhead Wilson,* identifying recidivist criminals, limiting immigration, readying the nation for war, or dealing with external and internal threats, systems of identification have long had many objectives. But whatever the objectives of the systems of identification, technical shortcomings have limited their effectiveness. Early on, the use of names and forms of identification by human sight quickly revealed the inadequacies of such systems. Manipulation, false representations, and misidentification drove the evolution of systems of identification, some more reliable and accurate than others. Bodily characteristics, which could not be altered, were the subject of interest for methods of identification not only in fictional accounts like *Pudd'nhead Wilson* but also in the work of morphological researchers in the 1800s. One of the earliest researchers on fingerprinting, Dr. Harris Hawthorne Wilder, described the problems with forms of identification based on names and residences:

> The people that pass in a double current upon either side of this stream of traffic are not thus registered; they bear no identification number, and have a place in the commonwealth simply by virtue of a personal name, recorded at the time of birth, and held simply in the memory of the individual himself and his personal acquaintances. Under numerous circumstances, some of them by no means rare, this loose system, relying as it does upon the individual memory, and the willingness to be identified, proves insufficient or actually misleading, and there is thus great need of a surer method of definitely describing and recording each human individual. This would seem possible under all circumstances only by making use of some mark or peculiarity permanently and unalterably fixed upon the body itself, and the best efforts of the anthropologists have now for many years been devoted to this question. (Wilder and Wentworth 1918, 17)

Wilder further explained that an obvious alternative to systems of identification based on name are the markings of nature inscribed on individuals, which had the potential to serve as an accurate form of identification:

> Humans are already marked by Nature, some months before birth, with a highly complex design, or system of design, unchanged throughout life, absolutely individual and impossible to duplicate, and for a number of reasons, remarkably resistant to decay. These markings are, besides, easy to record, and quite possible to formulate and classify, making it a simple matter to find a given record out of a set of many thousands within a few minutes. (26)

He then described a fundamental change taking place in systems of identification that gave rise to related and competing methodologies: fingerprinting and anthropometry.

The idea of using bodily characteristics for identification, despite their promised accuracy, has always presented moral and political questions. For as long as there have been systems of identification, there have also been persistent concerns regarding the consequences to those individuals who are identified. Conflicts with existing norms, values, and expectations, or the lack of legitimate purpose for a system of identification, have long challenged credentialing systems. The interests in controlling fraud, misrepresentation, and confirmation of identity have been present throughout the past two centuries, along with the potential threat of a system of identification to liberty interests. History thus lends insight to the continuing discussion.

The fluidity of the modern world and the increasing complexity of individual identity, coupled with the possibility of a wider variety of means of identity assurance, have intensified discussion of the appropriate scope of identification systems. Identity, wrote Marx (2001, 36), is "less unitary, homogenous, fixed and enduring as the modernist idea of being able to choose who we are continues to expand" (36). The need for identity assurance is obvious, but the problems associated with emerging technologies are apparent as well. New technologies in the public and private sectors present the capability to amass data for a myriad public and private functions, resulting in what Roger Clarke (1988) has termed "dataveillance." As a modern tool for identification, biometric technology serves the goal of identity assurance, but also triggers concerns regarding the misuse of data and the potential for violations.

The body has once again become a source as well as a site of surveillance. I say "once again," because there is nothing intrinsically new about the body being used in this way. Over a hundred years ago, criminal anthropometry claimed that body shapes, especially the head, could spontaneously reveal any unlawful proclivities of a person. Today's biometric technologies mean that the body itself can be directly scrutinized and interrogated as a provider of surveillance data (Lyon 2001).

Biometric technology thus serves the purpose of enhancing security but also provides opportunity for the violation of liberty. This is not a new problem for systems of identification, which have been long plagued by questions of legitimacy, accuracy, and reliability.

An Evolution of Identification

The rise of the modern state and society signaled the concurrent need to identify citizenry:

> Establishing the identity of individual people—as workers, taxpayers, conscripts, travelers, criminal suspects—is increasingly recognized as fundamental to the multiple operations of the state. . . . And not only the state: private economic and commercial activities would also grind to a halt unless companies had the ability

to identify and track individuals as property owners, employees, business partners, and customers. (Caplan and Torpey 2001, 1)

The most obvious solution to the difficulty of identification was to use what was already available: residence and name. In early-sixteenth-century France, for instance, the use of the patronymic was at first a social process tied to custom and then enforced by parish registers. Soon the Crown began to take over the registers as an exercise of authority over the Church and then later as a means of verifying the age of its ecclesiastical benefices. By 1667, the registers were an integral part of identifying individuals generally (Caplan 2001). The French Constitution of 1791 further defined the state's interest in lineage by establishing the legislative authority to "establish for all inhabitants, without distinction, the manner in which births, marriages, and deaths will be certified; and it will designate the public officials who will receive and maintain these new files" (Noiriel 2001, 29). When the Revolution in France brought attacks on the aristocracy, identified largely by their titles and names, a wave of name changing ensued, which led to legislation passed in 1794 forbidding the practice of changing names, thereby limiting the names used by individuals to those registered at birth.

The bureaucracies of the First French Republic (1792–1804) and the Napoleonic Empire (1804–1814) constructed individual identity not only with a name at birth, but also by collating available personal information, including addresses, ages, professions of parents, and witnesses when an individual was married. The collection of information around an individual's name and birthplace was an onerous administrative task fraught with human error. Misunderstandings due to dialects and differences of language created shortcomings in comprehension, which affected the consistency of the civil status record. The logistics of coherency were further undermined by the administrators tasked with creating the civil status system who tended to use marginal notes and inconsistent terminology that led to confusion and misinformation. There were also problems related to the naming of the Jewish citizenry where traditions were not accurately represented and, at the same time, the door was opened for the possibility of discrimination on the basis of naming. Notably, the apparent ease with which individuals could evade the system by giving erroneous information rendered this early attempt at systematic identification as flawed. The need for heightened accuracy and reliability increased interest in other forms of identification that made use of unique bodily characteristics: anthropometry and fingerprinting.

Anthropometry

In the 1880s, Alphonse Bertillon, a French law enforcement officer, began to develop a method of anthropometry to attempt to deal with the problems of deception that were endemic to a system of identification built on name and residence. In the preface of his book, he described the prospects for social stability if a reliable system of identification could be created:

> A very considerable portion of the crime and wrong which disturb the order of human society result either directly or indirectly from the apparent impossibility of distinguishing in every case and with unerring certainty one individual from another. It is for this reason, especially, that so many of the professional and habitual criminals who abound in every land have hitherto gone "unwhipt of justice."
>
> Men would be unlikely to render themselves liable to the penalties of the law if they knew that, wherever they might flee, their identity could not fail to be discovered. A sure means of identification would not only have the effect of deterring from crime in general, but would evidently nullify all attempts of whatever kind at a substitution of persons. No impersonations of a pensioner, or a missing heir, or a business man could ever hope to be successful. (Bertillon 1896, 1)

For Bertillon, the verification of identities was especially important to the penal laws of France, which were being undermined by criminals who concealed their identity. The answer to this practice was a system that Bertillon called "signalment" to both verify declared identity and reveal concealed identities. Bertillon set out to develop an anthropological method of description that was both reliable and adaptable to classification. The source of this system was found in the characteristics of the body. He wrote, "Nature never repeats herself. Select no matter what part of the human body, examine and compare it carefully in different subjects, and the more minute your examination is the more numerous the dissimilarities will appear; exterior variations, interior variations in the bony structure, the muscles, the tracing of the veins, physiological variations in the gait, the expression of the face, the actions and secretion of the organs, etc." (Bertillon 1896, 13).

Anthropometry was based on Bertillon's observation that human bone structure and skeletal development reached a level of stability and uniqueness during young adulthood and thus could be used to develop a system of description and classification. The Bertillon system was described as using a scientific approach of measurement and description:

The Bertillon system, although intended primarily for a practical end, can be made of scientific value as far as it goes. Its measurements are length and width of head, distance between zygomatic arches, length of left foot, of left middle finger, left little finger, left forearm, and length and width of ear. There is a descriptive part, including observation of the bodily shape and movements. Deformities, peculiar marks on the surface of the body resulting from disease or accident, and other signs, as moles, warts, scars, tattooings, etc., are noted. Experience has shown that absolute certainty of identity is possible by the Bertillon system.

The Bertillon system generated high expectations. Hope arose that a reliable system of identification would order society. Some argued, however, that universal application was necessary. In a report presented to the United States Congressional Committee on the Judiciary in 1902, the benefits of a general system of identification were set forth. But the full benefits of a practical system of identification cannot be reached unless applied to all individuals. There might be at first sentimental objections, as has happened in things subsequently of great utility to society. No one who intended to be an honorable citizen would have anything to fear; but, on the contrary, it would afford protection to humanity in enabling society to find its enemies. This certainty of identification would discourage dishonest voting, assist in recognizing deserters from the Army, in enforcing laws, and in facilitating many business matters. (MacDonald 1902, 999)

Although the expectations for the Bertillon system were high and the system gained official use in Russia, England, the United States, Belgium, and Switzerland, it was plagued with problems of reliability and accuracy (Wilder and Wentworth 1918).

The Bertillon system of identification was riddled with subjectivity arising from problems of training individuals in the accuracy of measurement, difficulties connected with instrumentation and calibrations, and failures to establish a positive identification system. When it became clear that the system was not adequate for positive identification, additional descriptive techniques were implemented in concert with anthropometry in order to allow the establishment of positive identity: the portrait of *parle*, or verbal portrait, which allowed the identification of escaped delinquents; the tabulation of distinctive marks to ensure forensic certainty; and the addition of photographs, which personalized anthropometric measurements (Kaluszynski 2001). These additional elements of description, however, also opened the door for more human error and inefficiency. As a result, in the first decade of the twentieth century, the use of anthropometric measurements declined, giving rise to increased use of fingerprinting.

Fingerprinting as a form of identification did not follow directly from Bertillon's system of identification and in fact developed alongside it. As early as 1858, William Herschel, as part of the small class of civil officials attempting to govern India, described the utility of fingerprinting.

Herschel, an East India Company administrator, was in charge of a subdivision in Jungipoor. During his magisterial tenure, he described his frustration with and distrust of all evidence tendered in courts and his revelation of a potential solution in the form of fingerprinting:

> I remember only too well writing in great despondency to one of the best and soberest-minded of my senior companions at Haileybury about my despair of any good coming from orders and decisions based on such slippery facts, and the comfort I found in his sensible reply. It happened, in July of that year, that I was starting the first bit of road metalling at Jungipoor, and invited tenders for a supply of "ghooting" (a good binding material for light roads). A native, named Rajyadhar Konai, of the village of Nista, came to terms with me, and at my desire drew up our agreement in his own hand, in true commercial style. He was about to sign it in the usual way, at the upper right-hand corner, when I stopped him in order to read it myself; and it then occurred to me to try an experiment by taking the stamp of his hand, by way of signature instead of writing. (Herschel 1916, 1)

By 1877 Herschel, who was then in charge of the Department for the Registration of Deeds in India, described the incorporation of two fingerprints on deeds and in a diary book as a means of verifying identity at a later date. This diary of fingerprints figured prominently in Francis Galton's work (Herschel 1916).

Galton, a cousin of Charles Darwin, was dedicated to exploring variation in the human population and its consequences. First driven to explore fingerprinting as a function of his interest in eugenics and heredity, Galton began to see it as a potential form of legal identification subject to the challenge of developing a system of classification:

> It became gradually clear that three facts had to be established before it would be possible to advocate the use of finger-prints for criminal or other investigations. First, it must be proved, not assumed, that the pattern of a finger-print is constant throughout life. Secondly, that the variety of patterns is really very great. Thirdly, that they admit of being so classified, or "lexiconised," that when a set of them is

submitted to an expert, it would be possible for him to tell, by reference to a suitable dictionary, or its equivalent, whether a similar set had been already registered. These things I did, but they required much labor. (Galton 1908, 254)

Even Bertillon, when contacted by Galton about the possibility of adding fingerprints to his system of identification, lamented that the practical difficulties of training in the methodology of fingerprinting would undermine its usability. When the Troup Committee was appointed in 1893 at the direction of the home secretary in England to debate the question of adopting fingerprinting or Bertillonage, they too decided that the classification system for fingerprinting was underdeveloped and rife with problems (Galton 1908).

Sir Edward Henry, a member of the British Indian police who later became the inspector general for Bengal in 1891, expanded and refined the beginnings of Galton's classification system. After extensive correspondence with Galton and a visit to Galton's laboratory in 1894, Henry devised a classification system for fingerprints that was then used in the Bengal police department. Galton's classification system of arches, loops, and whorls was expanded to include composites of whorls and subclassifications based on ridge tracing and ridge counting. This more intricate system also resulted in a more subtle classification system than Galton's. The development of a sound classification system allowed fingerprinting to overtake anthropometry in Britain. By 1900, when Henry published *The Classification and Uses of Finger Prints,* the Belper Committee in Britain, established in the same year, convened to resolve the contest between anthropometry and fingerprinting and decided to adopt fingerprinting on a trial basis (Cole 2001). In 1901, Henry became the assistant commissioner of police for criminal identification at New Scotland Yard, confirming the victory of fingerprinting in Britain (James 2005).

The acceptance of fingerprinting in the United States did not follow the same trajectory as the British experiment. Early uses of it in the United States did not originate in criminal justice as one might expect. Instead. use of fingerprinting as a system of identification first gained prevalence for limiting immigration in response to the trafficking of Chinese women for the purposes of prostitution in the 1880s.

The discovery of gold in 1848 brought thousands of immigrants from many countries to California. Mining was an exclusively male activity; few of the men brought families with them because, among other reasons, mining involved moving from place to place seeking the most productive site. Among the first female arrivals were prostitutes of varying racial and national origins. The demand for prostitution in San Francisco was partially met by Chinese women from Hong Kong, Canton, and its surrounding areas (Cheng 1984).

Initially, the American consulate in Hong Kong examined the Chinese women and made a determination of whether they were "lewd or debauched" before they were allowed to embark for America. If they were determined to be of good character, they were stamped on the arm and their photographs were taken and mailed to the collector of customs at San Francisco (Cheng 1984).

The need for a sound system of identification increased with the enactment of the Exclusion Act in 1882, which forbade the immigration of Chinese laborers for ten years. Although the legislation was designed to limit the immigration of laborers, a loophole allowed Chinese who were already living in the United States to leave and then return if they possessed a return certificate. Such a system was subject to the limitations of anthropometry description. The returning Chinese required identification so as to preclude the potential fraud of selling the return certificates. The first attempt to verify identities included a section on the certificate for description. "Clerks often squandered this single space by recording information of little descriptive value, such as 'flat nose,' 'large features,' 'small features,' or 'hole in right ear' (for an earring). This system was arbitrary and subject to individual whim and gave little hope of accurately identifying returning Chinese" (Cole 2001, 122). When the immigration of Chinese laborers was banned altogether by the Scott Act of 1888, the need for the return certificate was eliminated, and the use of fingerprinting ended.

Fingerprinting made a reappearance in 1903 as part of the New York civil service exams, when fingerprints were taken to discourage impersonation. At the same time, fingerprints were introduced as a form of identification in New York prisons (James 2005). Beginning in 1906, the U.S. Army also began to use fingerprinting as a way to identify deserters or discharged soldiers. Later, fingerprints were used as a way to identify the dead (Cole 2001).

By 1910, fingerprinting as a form of classification was used to fight the rise of prostitution in New York. The Inferior Criminal Courts Act mandated fingerprinting when prostitutes were brought into court (Cole 2001).

The attack on prostitutes was consistent with the trend of the era that gave a social nod to reverse eugenics, which advocated the removal of "genetic threats" from society. Progressive era reformers, who viewed prostituting as a sign of a biological flaw in certain women, required and, more important, substantiated the state's interest in identification of these women, who were considered to be a threat to social stability and morality. The belief that habitual criminality and "feeble-mindedness" had a genetic basis was condoned later by the U.S. Supreme Court in the now infamous 1927 decision in *Buck* v. *Bell*. Carrie Buck, whose mother was supposedly a prostitute and feeble-minded, was sterilized after the decision, in which Justice Oliver Wendell Holmes declared that "three generations of idiots are enough" (207).

The usefulness of fingerprinting for identifying prostitutes paved the way for its adoption in the identification of other habitual criminals by the New York Magistrate's Court, including men who had been arrested for "vagrancy and intoxication" (Cole 2001). Those who were identified as recidivists were distinguishable as a class in society. But a system of identification had even larger implications. The application of fingerprinting was described as a panacea for the ills of social disorder and the achievement of a better society. City magistrate Joseph Deuel credited social stability to the increased use of fingerprinting: "We are enjoying better order and decorum in the city than ever before, and much of the credit is due to the finger-print process" (Cole 2001, 158).

By the 1930s, the idea of combating crime with the use of a system of identification based on fingerprinting was well established. Driven by high-profile dramatic crimes, including the kidnapping of Charles Lindbergh's baby and Kansas City shootings involving Pretty Boy Floyd, a public awareness of crime and its threat was fanned , providing the grounds for the acceptance of fingerprint evidence when it was used to help solve these crimes. These crimes helped the efforts of J. Edgar Hoover to expand the National Fingerprint Bureau, established in 1924, later informing the U.S. House Appropriations Committee that he aspired to create a system of universal identification of every American citizen (Silver 1979).

Criticism served to kill the idea of universal fingerprinting or identification cards as public outcry against mandatory identification grew. Beyond the potential consequences for civil liberties, a universal form of identification also did not fare well because of the stigma of fingerprinting from its use in criminal identification. The early fear of universal fingerprint identification, even if it had been successfully implemented, would have been undermined by the difficulty posed by managing the growing numbers of fingerprints and the need to reduce the time required to capture multiple sets of prints. Technology eventually provided a solution to this problem. Digital scanning of fingerprinting could be done by 1982, but it was limited by the requirement that the finger be flat. Not until 1988 were rolled impressions digitalized. Finally, in 1995, the Federal Bureau of Investigation put technology in place that allowed electronic submission of fingerprints, leading to the creation of the Integrated Automated Fingerprint Identification Systems. The advent of digital processing paved the way for the evolution of different forms of systems of identification and new concerns about the use of systems of identification, like biometric technology.

Evolution of Biometric Technology

Biometric technology is often described as an extension of Galton and Bertillon's efforts to describe the unique characteristics of individuals combined with the arrival of digital processing. Digital signal processing techniques spurred interest in automating human identification in many biometric forms as the increased need and potential for developing high-security access control and securing transactions in both the public and private sectors rose. Factors of social acceptance nevertheless continue to be of concern. Social acceptance has been complicated by the variety of biometric technologies and the diversity of deployments. The physiological and behavioral attributes are ever evolving and include fingerprints, hand geometry, palm prints, vein, odor, voice pattern, iris, retina, face, ear, DNA, gait, signature, and keystroke.

Biometric identification systems using any of the many identifiers face challenges in achieving technological maturity and gaining societal acceptance. Societal perceptions of biometric identifiers vary according to the type of human interface required with the sensor at the point of collection, the quality of the biometric identifiers, the institutional objective, and the management of the database, among other issues. Despite these factors, at the most basic level, biometric technologies "are automated methods of verifying or recognizing the identity of a living person based on a physiological or behavioral characteristic" (Wayman et al. 2005, 1). Biometric technology uses a biometric pattern, physiological or behavioral, to discover the identity or verify the identity of an individual. For biometric identification systems, the question of maturity is not a straightforward matter but instead requires a mix of factors, including "robustness, distinctiveness, availability, accessibility, and acceptability" (Wayman et al. 2005, 3). Although biometric identification has been portrayed in the media as being able to seamlessly identify individuals without problem, a number of considerations, including societal acceptance, affect the successful deployment of biometric identification systems. Systems using any of the many identifiers face challenges in achieving technological maturity and gaining societal acceptance. Societal perceptions of biometric identifiers vary according to the type of human interface required with the sensor at the point of collection, the quality of the biometric identifiers, the institutional objective, and the management of the database, among other issues.

For biometric identification systems, the question of maturity is not a straightforward matter but instead encompasses a mix of factors, including "robustness, distinctiveness, availability, accessibility, and acceptability" (Wayman et al. 2005, 3). Although biometric identification has been portrayed by media as being able to seamlessly identify individuals without problem, many considerations affect the successful deployment of biometric identification systems, which necessarily includes a consideration of societal acceptance.

The task of identity assurance using biometric authentication is essentially one of pattern recognition that "extracts a salient feature set from the data, compares this feature set against the feature set(s) stored in the database, and executes an action based on the result of the comparison" (Jain, Flynn, and Ross 2008, 3). The process of biometric identification first requires that a physical characteristic or trait be captured with a sensor of some kind, generating a template that will then be used to authenticate the user's identity for physical access, distribution of benefits, or a variety of other objectives. In a one-to-one matching scenario, there is an enrollment phase and then a verification phase. The system confirms an individual's identity by comparing it to that individual's existing biometric template from a database of templates of a limited universe of individuals (Jain 2003). For instance, a number of states have used biometric identification to combat fraud in welfare payments. Beginning in 1994, Los Angeles County in California teamed with Printrak to provide the Automated Fingerprint Image Reporting and Match (AFIRM) system. When a new welfare applicant applies for benefits, a fingerprint of the forefinger of each applicant's hand is taken. The personal information of the applicant is stored in a database and linked to a data storage subsystem that holds the fingerprint image data.

In a one-to-many matching scenario, a biometric identification system is operating in a discovery of identity mode when it recognizes an individual by comparing the template against users in one or more databases (Jain, Ross, and Prabhakar 2004). A one-to-many matching scenario can amplify some of the difficulties of establishing an individual's identity.

After the sensor captures the biometric data, the processes of quality assessment and feature extraction follow. Signal processing is a class of algorithms used to remove irrelevant noise from data or to help highlight important features and create a template (Woodward, Orlans, and Higgens 2003). Segmentation, in which the relevant biometric data are separated from background information, is followed by a process of feature extraction. Depending on the type of identifier in use, the biometric pattern must be distinguished from the distorting qualities by a process of extraction. "In a text-independent speaker recognition system, for instance, we may want to find the features, such as the mathematical frequency relationships between vowels, that depend only upon the speaker, and not upon the words being spoken, the health status of the speaker, or the speed, volume and pitch of the speech" (Wayman et al. 2005, 12).

The extracted features are compared against the stored templates to generate match scores to either verify identity or discover identity. In this process, there can be errors, including failure to capture (FTC) and failure to enroll (FTE). The FTC rate can be relevant when the biometric identification system is used in an open event or in an automatic enrollment process. This type of error occurs when the biometric identifier is lacking in quality—an extremely faint fingerprint or an occluded face, for example. This is especially true in a large-scale screening process because the accuracy of biometric identifiers can be compromised by environmental factors. The FTE rate denotes the percentage of instances in which users cannot enroll in the recognition system. Ideally, the templates are high quality and rejection rates on this basis are low. FTE errors typically occur when the system rejects poor-quality templates during enrollment.

Other types of errors that can occur are false acceptance rates (FAR) and false rejection rates (FRR). In the FRR case, an individual who should be positively identified is rejected erroneously, while in the FAR case, the subject is identified incorrectly. A false match (FMR) occurs when biometric measurements from two different persons are read to be from the same person. A false nonmatch (FNMR) occurs when two biometric measures from the same person are read to be from two different individuals. (Jain 2003). These two points of technical fallibility are implicated differently depending on the context of use. As Jain, Ross, and Probhakar explain, "in some forensic applications, such as criminal identification, the FNMR rate (not FMR) is the critical design issue: that is, we do not want to miss a criminal even at the risk of manually examining a large number of potentially incorrect matches that the biometric system identifies. On the other extreme, FMR might be one of the most important factors in a highly secure access-control application, where the primary objective is deterring impostors" (Jain, Ross, and Prabhakar 2004, 7)

Like the systems of identification that preceded them, biometric identification systems also suffer from the problem of fictional versus factual identification when biometric identifiers are compromised. A fingerprint that is damaged or scarred, a voice that is affected by a cold, or a face that is changed by cosmetic surgery can affect the accuracy and reliability of biometric identifiers. Another issue that can arise with a biometric identification system is intraclass variation. When enrollment data do not match the characteristic during the verification phase, the system will have difficulty verifying an identity. These types of variations might be differences in signature, gait, or voice recognition because of physiological or behavioral changes. A corollary of the problem of variation is distinctiveness. Among individuals, there may be substantial similarities in a biometric trait that may result in a limitation on the capability to distinguish among individuals. Another one of the more popularized circumventions of a biometric identification system is a spoof attack: the attempt on the part of an individual to trick the system by assuming the biometric trait of another individual. This type of attack is thought to be more common with behavioral characteristics that can be adopted more easily but may also apply to the fraudulent creation of a "spoofed" fingerprint by creating a model of a latent capture of a fingerprint. These technological fallibilities, however, are not the only obstacles that biometric identification systems must overcome. Societal acceptance is key factor in evaluating the eventual success of any modern form of identification.

Social Acceptance and Modern Forms of Identification

Biometric identification systems are obviously not the first form of identity assurance. Modern forms of identification have greatly expanded in the context of the information revolution for which identity assurance is a growing concern. The occurrences of identity theft and large-scale losses have prompted the wide use of systems of identification by the public and private sectors, but there is still a dominant concern of social acceptability when a system of identification is chosen.

The societal acceptance of a system of identification is a complicated calculation. Consider the controversy surrounding the use of the social security number. The Social Security Act (1935) raised concerns on both sides of the political aisle. The United Steel Workers and the United Mine Workers were concerned about the capabilities of the card, fearing the possibility of being blacklisted (Parenti 2004). This sentiment was so widespread among the American public that the drafters of the act avoided the word *registration* because of the possibility that the entire idea would be rejected. Also of concern was the institutional responsibility for the assignment of numbers. Arthur Altmeyer, the acting chair of the Social Security Board, asked the postmaster general to handle the issuance of the social security numbers because of the view that Americans trusted the post office more than the rest of the government (Watner and McElroy 2004). When President Roosevelt signed Executive Order 9397, the groundwork for future expansion of the social security number as a personal identifier beyond the social security program was laid with the order that all federal agencies use the number whenever it was advisable to set up a new identification system for individuals. Despite the enabling language of Executive Order 9397, the use of social security numbers was limited until the arrival of computer systems, when the utility of the number exploded and governmental agencies began to use it as a form of identification and individuals self-identified with it (Watner and McElroy 2004).

In time, the use of the social security number spread through the public and private sectors, giving rise to the misuse of the number for fraud and misrepresentation, which led to considerations of more secure systems of identification. Traditional identification information, including mother's maiden name, social security number, and drivers' license number, have been used to shore up identity assurance, but success has been limited, and confidence in them is waning. This lack of confidence in personal identifiers can affect the currency of information for which our transactional, communication, and social networking relationships are increasingly used. A modern form of identification is only as reliable as the type of information it is built on, making it necessary to assess how different forms of personal information compare. If certain types of personal information are viewed as safer for identity authentication, greater confidence in these identifiers will facilitate interactions. In this equation, safety is indicated by familiarity, ease of replication, consistency, uniqueness, and readability.

Assessing Validity and Reliability

To establish an evaluative context for biometric identifiers, respondents in the focus groups and survey in the research reported in this book were asked to consider the validity and reliability of various forms of personal identification: mother's maiden name, social security number, bank account number, and date of birth, and, for comparison, several biometric identifiers. The focus group participants, when asked about the safety of a variety of identifiers of personal information, rated fingerprint and voice recognition as safer than other identifiers. Many rated user name and password and voice authentication as somewhat safe. Social security number, credit card number, date of birth, driver's license number, and address were generally perceived as less safe. The only major difference between nonusers of biometric technology and user focus groups was the perceived safety afforded by the use of social security number and driver's license number as a form of identity authentication. Nonusers generally ranked social security number as "not at all safe" and driver's license

Table 1.1
Average ratings assigned by biometric users and nonusers: "how safe do you feel each of the following types of information is as a way to protect your personal records from access unauthorized persons?"

Information type	Biometric user mean (sd)	Biometric nonuser mean (sd)	Total mean (sd)
Social security number	2.67 (1.48)	1.78 (1.10)	2.34 (1.40)
Driver's license number	2.84 (1.27)	2.10 (1.17)	2.56 (1.28)
Mother's maiden name	2.67 (1.38)	3.10 (1.46)	2.81 (1.41)
Date of birth	1.81 (1.05)	1.88 (0.98)	1.84 (1.02)
Place of birth	2.15 (1.15)	2.81 (1.38)	2.40 (1.27)
Address	1.47 (0.79)	1.72 (0.92)	2.40 (1.27)
Credit card number	2.33 (1.19)	1.81 (1.01)	2.14 (1.15)
Bank account number	2.73 (1.25)	2.16 (1.11)	2.52 (1.23)
Fingerprint	4.58 (0.79)	4.17 (1.37)	4.44 (1.04)
Voice authentication by phone	3.76 (1.05)	3.40 (1.19)	3.67 (1.09)
User name and password	3.11 (1.13)	3.03 (1.27)	3.08 (1.18)
Home telephone number recognized by an automated system when you call in from that number	2.64 (1.27)	2.31 (1.35)	2.53 (1.3)
Group means	2.73	2.52	2.66

Note: Ratings were collected using a Likert scale from 1 to 5: 1 = Not at all safe, 2 = Somewhat unsafe, 3 = Neither safe nor unsafe, 4 = Somewhat safe, and 5 = Very safe.

number as "somewhat unsafe," while users seemed to average both as "somewhat unsafe." (See table 1.1.)

The uniqueness of a piece of identifying information and the difficulty of duplicating the information were critical to how participants ranked the overall safety of the identifiers. Participants reported mixed views of the traditional identifiers (e.g., social security number, drivers' license, place of birth), citing wide availability and use as undermining their safety as identifiers. A minority of participants stated that a user name and password were safe because they had some degree of control over them and could change them if a privacy violation occurred. Some participants rated mother's maiden name and place of birth as relatively safe, because these pieces of information are not widely known by others and do not tend to be listed or recorded in places other than a birth certificate. Nonetheless, many participants across groups suggested that if the information can be stolen, overall level of security falls and can result in a violation of informational privacy. Generally address, date of birth, and drivers' license number were perceived as less safe. Participants reportedly feared that these identifiers could be easily accessed by thieves and hackers.

Results from the national survey reflected similar views. Respondents were asked to indicate how safe they felt when using a variety of biometric and nonbiometric identifiers, using a scale of 1 to 5, with 1 representing "not at all safe" and 5 representing "very safe." The respondents' ratings, shown in table 1.2, indicated that they felt the safest using fingerprints, their mother's maiden name, their birthplace, or a user name and password.

Respondents' level of education influenced their perceptions of the safety of identifiers; the analysis provided evidence that the difference in

Table 1.2
Average reported levels of safety of various identifiers ($N = 1{,}000$)

Identifiers	Mean (sd)
Social security number	2.6 (1.4)
Driver's license number	2.7 (1.4)
Mother's maiden name	3.2 (1.4)
Date of birth	2.8 (1.4)
Place of birth	3.2 (1.4)
Street address	2.5 (1.4)
City	2.6 (1.5)
Country	2.7 (1.6)
Zip code	2.7 (1.5)
Credit card number	2.5 (1.5)
Credit card expiration date	2.7 (1.4)
Bank account number	2.7 (1.5)
Fingerprint	3.8 (1.5)
Phone number	2.3 (1.4)
Fax number	2.3 (1.4)
E-mail address	2.3 (1.3)
Occupation	2.6 (1.4)
User name and password	3.0 (1.5)
Home telephone number recognized by an automated security system when you call in from that number	2.7 (1.4)

Table 1.3
Impact of age on reported levels of safety with various identifiers (N = 1,000)

Identifiers	Correlation coefficients
Mother's maiden name	$r = .11, p = .001$
Street address	$r = .12, p < .0001$
City	$r = .13, p < .0001$
Country	$r = .11, p = .001$

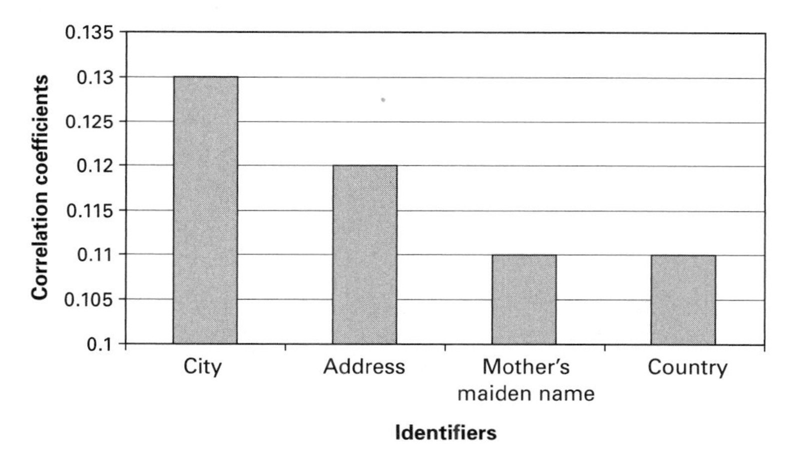

Figure 1.1
How age impacts reported levels of safety with various identifiers

education levels was statistically significant. Individuals with a high school diploma or GED perceived a higher level of safety with certain identifiers than those with some college or more. (For how education affected perceptions of various forms of personal identifiers, see appendix A.)

There were also statistically significant, positive correlations between age and reported levels of safety with other identifiers. Table 1.3 and figure 1.1 present the Pearson correlation coefficients for each variable.

Safety of Different Biometric Technologies

The perceptions of the safety of biometric technologies were also assessed for comparison purposes. Social acceptability hinges on many factors. There are obvious differences between public and private sector applications and mandated versus optional systems, but there are also differences that are made significant depending on the societal perception of the biometric identifier in question and the purpose to which it is put. Some identifiers may be more socially acceptable but suffer from inconvenience or a lack of reliability. Others are convenient and unobtrusive and thus win societal acceptance even if they suffer from concerns about safety. It is also important to realize that societal perceptions do not always accurately affect the true status of technological maturity or precisely represent the potential pitfalls of interacting with the technology. Opinions of researchers involved in the minutia of biometric testing and assessment with a sophisticated understanding of how each biometric technology works do not serve as an accurate barometer of social acceptance. Societal perceptions, even erroneous ones, will continue to dictate the success of biometric technology. Societal perceptions may be affected by various factors, including the type of interface with the sensor, perceptions of quality, and method of collection.

For example, depending on the type of biometric reader used, there are different requirements of human interface with the sensor. The type of interface between the sensor and the individual can influence societal acceptance. When the sensor is remote and the individual does not give consent, societal distrust can arise. Facial recognition technology generated societal concerns when it was deployed to surveille larger public spaces to combat crime or detect potential terrorists. Although constitutional protections clearly do not safeguard those things willingly revealed in public, the societal acceptance of surveillance is not always aligned with the law. When facial recognition technology was installed in Ybor, a neighborhood in Tampa, Florida, shortly after September 11, 2001, it was met with a barrage of criticism and was eventually abandoned. The reason was lack of societal acceptance. While police spokesman Joe Durkin told the *St. Petersburg Times,* "I wouldn't call it a failure," the company providing the technology, Identix, issued this one-sentence statement: "Identix has always stated that this technology requires safeguards, and that as a society, we need to be comfortable with its use" ("Big Blind Brother" 2003).

When the interface with the sensor might be perceived as invasive to bodily integrity, it can also affect societal acceptance. Retinal scans have suffered criticism because of the societal belief that this type of biometric identifier is akin to angiography, which captures dye-enhanced images over time to investigate blood circulation in a medical procedure. In truth, retinal scans for the purposes of biometric identification make use of only a few images and are focused only on a small portion of the retina. However, despite the fact that the societal perception might be erroneous, the belief that a certain kind of biometric identifier is more invasive or dangerous than others will continue to affect its relative success because it is viewed as invasive to bodily integrity.

Societal perceptions of biometric identifier quality also might undermine social acceptability. The use of hand geometry has been popular for physical access because of its simplicity, ease of use, and low cost; however, difficulty arises when a hand geometry verification system is used on large populations because the geometry of the hand is not as distinctive as some of the other forms of physiological identifiers. The shape and distinction of the hand can be changed by growth, age, or jewelry (Biometrics.gov, 2007). Hand geometry might be acceptable in one-to-one matching circumstances but socially unacceptable when it is used in a one-to-many matching situation because of problems of quality.

These types of quality concerns also affect voice recognition technology. Voice is both a physiological and a behavioral biometric identifier. Although the physiological characteristics are invariant, changes can occur over time because of colds, age, or deception. Three types of applications are usually associated with voice biometrics: voice authentication, usually for access control; speaker detection, used in call centers or for wiretapping or surveillance; and forensic speaker recognition, for use as evidence in courts or for intelligence (Jain, Flynn, and Ross 2008). For large-scale identification, voice recognition is not as accurate or reliable as some of the other biometric identifiers, and background noises can affect the process of verification (Jain, Ross, and Prabhakar 2004). Yet despite these quality concerns, voice authentication has gained prominence in the financial institution sector. A Frost and Sullivan report estimated the growth of the global biometric voice verification market to be rapid between 2004 and 2011, with a compound annual growth rate of 49.8 percent during this period. An increased adoption of voice verification solutions by financial institutions is expected to be a primary driver for this growth. The attractiveness of voice recognition for the financial industry lies in its convenience, efficiency, and customer acceptance. For these reasons, voice verification is often used in concert with other forms of identification, incorporating a password with a voice biometric. The quality concerns that might have affected societal perceptions are mitigated by other advantages of this institutional use. Voice verification may not, in other words, be socially acceptable if it is used in a one-to-many situation.

Another way in which quality is a factor in societal acceptance is found in the perceived differences between behavioral biometrics and physiological types of biometrics. Although behavioral forms of biometrics are thought to be less reliable than some of the physiological biometric characteristics, they are often favored because of their lack of intrusiveness and are often widely used in concert with other forms of identification. For some of the behavioral biometric identification systems, convenience, efficiency, and unobtrusive collection do not always translate into societal acceptance. Concerns of quality continue to predominate. Gait recognition is one example. Although this biometric identifier can be collected unobtrusively, it suffers from the assumption that it is not unique and therefore cannot provide highly accurate identity verification, in a one-to-many matching situation. Recent innovations in gait recognition technology, however, might be changing the concerns about quality. New systems use a camera with a side view to record a set of key frames, or stances, as an individual completes a full walk cycle. These frames can then be converted into silhouette form and analyzed with height measurements and the periodicity of the gait to identify the person more accurately than in the past.

Keystroke identification is another form of behavioral biometric that might be perceived as lower in quality, but it is gaining popularity in certain sectors because of its limited institutional objective and because of considerations of convenience, efficiency, and unobtrusive collection. Although one individual may have a wide range of variations in his or her typing patterns, there is sufficient discriminatory information association with keystroke recognition to use with other forms of identification. In a testing commissioned by BioPassword(software that learns and verifies unique typing patterns). the Tolly Group, an independent information technology test lab, found that the software thwarted 99.2 percent of its fraudulent log-in attempts and allowed 98 percent of legitimate log-ins, which addresses concern over false positives (Roiter 2008). This trend is similar with regard to signature biometric systems, which are in widespread use in government, legal, and commercial transactions. Issues that arise with signature biometrics include the potential for change over time and the influence of a subject's physical and emotional conditions (Jain, Ross, and Prabhakar 2004). Nevertheless, growing demand from the governmental, financial, and health care sectors to prevent check fraud and reduce costs associated with paper-based documents has expanded the potential market for digital signature verification (DSV), earning $14.4 million in 2006 and potentially reaching $85.7 million by 2013. Frost and Sullivan research analyst Neelima Sagar described the future of signature verification: "High social acceptance and convenience of DSV technologies will ensure that the market continues to witness substantial sales." Issues of the quality of signature recognition have been mitigated by limited institutional use, convenience, and efficiency.

Rating Biometric Identifiers

To build an understanding of societal perceptions of different types of biometric identifiers, focus group participants in the research for this book were asked to rate the safety of eight types of biometric information systems (see table 1.4). Across all groups, participants rated fingerprint and iris and retinal scans as the safest overall. They were reluctant, however, to suggest that biometric technologies they knew little about were safe measures.

After the initial rating, nonusers were given a short description of how the biometric identification system worked. They also were told that the data stored would not be sufficient to recreate the identifier itself. With this additional information, views on safety tended to rise. After hearing this information, participants altered their ratings in the following ways:

- The ratings for safety of voice recognition climbed substantially.
- Participants tended to rate voice recognition as highly as they rated iris and retinal scans and fingerprints.
- The safety ratings for fingerprints and face and hand recognition rose.

In discussions that followed, participants commented that they considered fingerprints to be safer than other pieces of information due to their

Table 1.4
Average ratings assigned by biometric users and nonusers: "How safe do you feel each of the following types of biometrics is as a way to protect your personal records from access by unauthorized persons?"

Type of biometric	Biometric user mean (sd)	Biometric nonuser mean (sd)	Total mean (sd)
Facial recognition	3.60 (1.13)	3.93 (1.06)	3.71 (1.12)
Hand geometry	3.76 (1.06)	3.92 (1.21)	3.82 (1.11)
Gait recognition	2.45 (1.10)	2.78 (1.28)	2.56 (1.17)
Voice recognition	3.60 (1.05)	3.27 (1.26)	3.48 (1.13)
Fingerprints	4.45 (0.86)	4.31 (0.90)	4.40 (0.87)
Key stroke recognition	2.53 (1.25)	2.27 (1.40)	2.44 (1.29)
Signature recognition	2.20 (1.18)	3.13 (1.30)	2.53 (1.29)
Iri/retinal scans	4.45 (0.83)	4.68 (0.54)	4.53 (0.75)
Group means	**3.38 (1.06)**	**3.54 (1.12)**	**3.43 (1.10)**

Note: Ratings were collected using a Likert scale from 1 to 5. 1 = Not at all safe, 2 = Somewhat unsafe, 3 = Neither safe nor unsafe, 4 = Somewhat safe, and 5 = Very safe.

uniqueness. One participant commented, "I think for every person it's very rare . . . that two people would have the same fingerprint."

For a minority of the respondents, fingerprint technology, which received a high rating for safety, nevertheless prompted concerns about reliability. This perception seemed to be precipitated by the common knowledge regarding the ways in which fingerprints can be lifted and some confusion between biometric identification using fingerprints and forensic identification. There was concern that fingerprints were not entirely safe, with some noting that they are left everywhere and could be obtained from any object a person had touched. Largely, this view was informed by media portrayals of fingerprinting, including crime shows such as *CSI*. The fact that fingerprints could be easily obtained gave rise to the perception that the use of fingerprints as an identifier could become troublesome if their use became widespread. Not only was the ease of gathering fingerprints a threat to verification and identification, but some users of fingerprint recognition suggested that fingerprint readers are not perfect and can sometimes be manipulated. One user of fingerprint recognition said that at his workplace, the fingerprint reader does not work when the weather is cold. Two others reported these views: "At work] we have a system . . . that I've beat by smooshing my thumb print," and "Half of the time I can't get in with my own fingerprint." Users of fingerprint technology, however, indicated that they understood how the biometrics work and that when the technology works correctly, it is a safe form of security.

A minority of nonusers expressed skepticism about the "reading" of biometrics. For example, one participant noted that interpreting fingerprints takes an expert and observed that automated telephone systems have mixed success at interpreting even what word a caller is saying.

The greatest difference between users and nonusers was the perceived safety of signature recognition. Nonuser responses averaged a rating of "fairly neutral," while users averaged "somewhat unsafe." Across nonuser and user groups, gait, keystroke, and signature recognition were rated lower than the other technologies. Also, some nonusers expressed initial skepticism about the reliability of voice recognition biometrics, noting, for example, that voices can be altered by illness. In addition, a few participants stated that features that can be observed in public were less safe because they might be stolen in some way. One nonuser, for example, referred to the fact that facial features are public and that a picture could be taken of a person unobtrusively and used for facial recognition.

Participants in the focus groups viewed technologies as safest when they were harder to reproduce, consistent, difficult to manipulate, unique to the individual, "readable," and less visible. Discussion across all groups revealed that in some cases, certain technologies were viewed as easier to copy than others. For example, some participants stated that it would be easier to mimic signatures or keystrokes than to find a substitute for a person's hands or eyes. They noted that some features of a face cannot be changed, making facial recognition a safer technology. Participants also often referred to the likelihood or unlikelihood that more than one individual would have the same biometric feature. Features that were strongly perceived as unique were often likely to be viewed as safer.

Another criterion for the safety ratings that emerged in discussions was the participants' perceptions of how consistent an identifier would be from day to day. Voice, the way in which people walk, and the way people type were all viewed as open to variation. For example, one nonuser commented

Table 1.5
Reported levels of safety using various biometrics (N = 1,000)

Type of biometric	Mean (sd)
Facial recognition	3.5 (1.3)
Hand geometry	3.5 (1.3)
Gait recognition	2.7 (1.2)
Voice recognition	3.3 (1.2)
Fingerprint recognition	4.1 (1.1)
Keystroke recognition	2.6 (1.3)
Signature recognition[a]	2.9 (1.3)
Iris or retinal recognition[b]	3.9 (1.2)

[a]There was a statistically significant difference in respondents' perceptions of level of safety. Individuals with some high school reported a higher perception of level of safety than those with some college or a four-year college degree ($F_{(df=5)}$ = 3.21, p = .007; Some high school: mean = 3.43, sd = 1.46; four-year degree: mean = 2.76, sd = 1.13).
[b]There was a statistically significant negative correlation between age and reported level of safety with iris/retinal recognition (r = –.16, p = .001). This suggests a contradiction of the original focus group findings that revealed that older nonbiometric users rated the safety of the technologies higher than other participants did.

that her gait depended on which shoes she was wearing. A user wondered how reliable gait recognition would be if a person hurt an ankle. Many participants across groups wondered whether such behavioral biometrics would be interpreted consistently. Despite these concerns, biometric identifiers were rated as significantly safer than the scores given the other forms of identifiers.

This finding was replicated in the national survey, where respondents were asked to indicate the perceived safety level of various biometric identifiers on a scale of 1 to 5, with 1 representing "not at all safe" and 5 representing "very safe." As shown in table 1.5, the respondents' ratings matched the focus group participants' comments and indicated that they felt the safest using fingerprint and iris or retinal recognition, followed closely by facial recognition and hand geometry.

The findings of the focus groups and the national survey are similar in the strength of the identity assurance assumed with biometric identifiers, though indicating a slightly lower level of acceptance for behavioral identifiers. Although the data reveal that biometric identifiers might be a "safer" alternative to other forms of identifiers, many concerns continue to prevail. Societal acceptance is also determined by implied consent and considerations of a limited population. Consider the US-VISIT program (United States Visitor and Immigrant Status Indicator Technology). When a foreign traveler obtains a visa, fingerprints and a photograph are taken and then checked against a database of known criminals and suspected terrorists. The interaction with the sensor is a necessary prerequisite to entering the United States and in this respect is considered implied consent; the requirement affects a specifically defined class. It is not, in other words, generalized surveillance.

Consent or implied consent is an important factor when considering the acceptability of a human interface with a biometric identification system. When Disney used fingerprint and then hand geometry to avoid the subjectivity of using a photo identification to confirm the identity of a ticket purchaser at its parks, criticisms were largely quieted by the fact that participation in the program was consensual. This was also true in the INPASS program, a voluntary program instituted by the U.S. government that has proven successful for frequent travelers, which involves a live scan of the bone structure in the hand to an image that was taken during the application for the pass. Other popular uses of hand geometry in the public and private sectors are access control to verify the identities of individuals seeking access to controlled areas, time and attendance verification, and metering resource use. These applications of biometric identification systems are not, of course, entirely without some constraints on individual choice. As a requirement of employment, access to Disney, or permission to enter the United States, the individual is not wholly free to choose to use the biometric identification system used, a concern that arose in the focus group discussions.

Several participants expressed strong opposition to the possibility that biometric information could be collected about them without their knowledge, while some users noted that signing up for the biometric was not troubling at all. One user said, "I recently had to go through the whole fingerprinting process. I actually felt very uncomfortable with it; I didn't feel like it was really necessary. . . . I felt they were gathering unnecessary information on me." Another commented: "When I first went to my company, they had biometrics. . . . I refused to do it. There was way too much information out there about me. . . . I have become more sensitive about data gathering. . . . I don't want it to be a fact of life."

Issues of societal acceptance do not end at the point of collection. After the collection of biometric identifiers, the data are deposited into a database. This process can include combining the identifier with other pieces of information or may involve the categorization of individuals, as in a no-fly list or on a terrorist watch list. Technological or human error can result in inadvertent disclosure, exposure, or distortion and can affect consideration of social acceptance.

The possibilities for human error are immense. For example, about seventy countries, including the United States, now use the epassport, a new standard of border control containing a biometric identifier. To many, this represents an advance in border security, but the level of security depends on the officer at the border matching the passport to the passenger at the counter. In a report released in March 2009 by the Government Accountability Office, a congressional investigative service, investigators managed to obtain four epassports using fake IDs because officials did not verify the underlying documents submitted for the passport (Temple-Raston 2009).

Potential vulnerabilities are also exacerbated by unknown secondary uses of biometric identifiers, or mission creep, when the information is gathered for one purpose but is shared across institutional boundaries for an objective other than the one acknowledged by the individual. This problem is associated not only with biometric identifiers. An example of this type of violation occurred with the clientele of the San Francisco Planned Parenthood organization. When the Department of Justice subpoenaed abortion records from Planned Parenthood in an attempt to compile information related to the recently passed Partial Birth Abortion Act, a problem of secondary use and, consequently a violation, occurred. Aggregating data raises the specter of the information being used for purposes that were not originally intended, thwarting an individual's freedom to make intimate decisions without governmental intrusion. Although this case involved confidential medical records explicitly tied to an individual's decisions to have an abortion, biometric information could also be used in situations not anticipated in the initial collection phase. Consider that national identification cards in the United Kingdom, necessary for registration for health

Table 1.6
Reported levels of concern with biometric use (N = 1,000)

Statements	Mean (sd)
That government agencies might use biometrics to track their activities[a]	3.2 (1.5)
That private industry might use biometrics to track their activities	3.5 (1.4)
That their biometric information might be stolen	3.5 (1.5)

[a]There was a statistically significant difference in respondents' level of concern with government agencies using biometrics to track their activities. Democrats were more concerned than Republicans ($F_{(df=4)} = 4.09$, $p = .001$; Democrats: mean = 3.50, sd = 1.43; Republicans: mean = 3.00, sd = 1.49).

and child care, welfare payments, and overseas travel, also include biometric identifiers that link to a range of databases, such as the National Identity Register, a national database of fingerprints (Rule, 2007). Biometric identifiers may serve a range of institutional objectives, many of which may or may not accord with the original intention of the individual.

These kinds of concerns were amplified in the national survey, when respondents were asked to indicate their level of concern with threestatements regarding potential negative uses of biometrics. The respondents articulated their concern on a scale of 1 to 5, with 1 representing "not at all concerned" and 5 representing "very concerned" (table 1.6).

The Currency of Personal Information and Safety of Identifiers

As these findings reveal, certain types of biometric identifiers have higher safety ratings as compared to other forms of identity assurance, but it does not follow that their role in enhancing the currency of personal information in social interactions in the public and private sector is secured and wholly accepted by the public. Fears of fallibilities and misuse of biometric identifiers continue to persist and complicate the deployment of identification systems, and they must be fully understood if biometric identification systems are to be accepted. Biometric technology as a means of identity assurance can serve an important role in social interactions increasingly built on the currency of personal information, but technological maturity must include a consideration of the societal reactions. The question then is how to use biometric systems of identification to facilitate the use of the currency of personal information to ensure identity in public and private transactions or in pursuit of public policy goals such as preventing terrorism or reducing the threat of identity theft. Although the societal impressions of biometric identifiers in comparison with other types of identifiers indicate a predisposition to believe that biometric identifiers are accurate and reliable, this is only the first step. The next step is to better understand the elements that give rise to the societal acceptance of systems of identification.

Public Support of Systems of Identification

Societal acceptance or rejection can shore up a system of identification or destroy it. The societal assessment of the objectives of a system of identification in the public or private sector is no easy matter, yet it is safe to say that ignoring the political, societal, and cultural influences that shape perceptions of systems of identification is impossible. It is clear that the influences of September 11 and increasing reliance on personal information figure into the dynamics of societal acceptance of biometric technology. Many have argued that September 11, coupled with the growing prevalence of information technology, provided a window of opportunity for the societal acceptance of biometric technology, leading the public to set aside concerns of technological surveillance in favor of achieving, even if imperfect, pie-in-the-sky security. Jeffrey Rosen (2005) contended for instance that "before 9/11, the idea that Americans would voluntarily agree to live their lives under the gaze of a network of biometric surveillance cameras, peering at them in government buildings, shopping malls, subways, and stadiums, would have seemed unthinkable, a dystopian fantasy of a society that had surrendered privacy and anonymity" (34). Nevertheless, there were signs that traditional ideas were being transformed with the use of a currency of personal information to network, carry out transactions, and communicate long before September 11. The issue, as Solove (2004) argued, is that "unless we live as hermits, there is no way to exist in modern society without leaving information traces wherever we go. . . . Privacy involves an expectation of a certain degree of accessibility of information" (143).

Between these two extremes is the truth about the impact of September 11 on societal acceptance of biometric systems of identification. Is our interest in individual liberty eroding in a post–September 11 world, or does it require different forms of technological and policy protection in a wired world? To answer this question, it necessary to examine the impact of September 11 on societal acceptance of biometric technology and consider whether the event translated into the surrender of our individual liberty to the tools of technological identification and surveillance.

2 September 11: A Catalyst for Biometrics?

An anniversary can be sweet or solemn, but either way, it is only the echo, not the cry. From this distance, we can hear whatever we are listening for. We can argue that September 11 changed everything—or nothing.

—N. Gibbs et al., *What a Difference a Year Makes* (2002)

Very deep is the well of the past. Should we not call it bottomless? . . . For the deeper we sound, the further down into the lower world of the past we probe and press, the more do we find that the earliest foundations of humanity, its history and culture, reveal themselves unfathomable. No matter to what hazardous lengths we let out our line they still withdraw again, and further, into the depths. *Again* and *further* are the right words, for the unresearchable plays a kind of mocking game with our researching ardours; it offers apparent holds and goals, behind which, when we have gained them, new reaches of the past still open out—as happens to the coastwise voyager, who finds no end to his journey, for behind each headland of clayey dune he conquers, fresh headlands and new distances lure him on.

—Thomas Mann, *Masks of Gods*, 1959

September 11 and the Information Revolution

While September 11 heightened the controversy over biometric identification systems, the increasing reliance on the currency of personal information had already set in motion the grounds for public awareness of the consequences of an increasing presence of information technology in our day-to-day lives. Rule (2007) aptly described this:

Organizations are constantly finding new ways of capturing, transmitting, and using personal data, for purposes defined by the organizations rather than those depicted in the data. Supermarkets track our purchases; government agencies track our travels, transactions, communications, and associations; insurers and employers monitor our medical histories and genetic makeup; retailers monitor our expenditures, our website visits, and our financial situations. (13)

In this manner, systems of identification were becoming part and parcel of the social fabric long before biometric identification systems gained notoriety after September 11. The specter of September 11 burdened biometric technology with expectations of security and criticisms about loss of privacy and civil liberties. However, societal perceptions of biometric technology were not so black and white, nor were they constructed tabula rasa. Societal perceptions of biometric technology reflect not only the current imperatives of identity assurance, but also the desire to maintain enduring normative values. It is helpful to understand the influence of September 11 on the political debate surrounding biometric technology by looking to the past.

Although the attacks of September 11 were unique, the aftermath was reminiscent of other times in U.S. political history when external and internal threats prompted governmental efforts to turn their attention to security. Not surprisingly, the quest for security has also raised concerns about the protection of civil liberties at other times. For example, the Alien and Sedition Acts of 1798 created presidential power to deport suspicious immigrants.[1] The acts, which also made it a crime to publish defamatory material against the Congress, the president, or the U.S. government, began one of the first serious discussions about the reach of First Amendment protections for the guarantee of freedom of the press. The Alien and Sedition Acts were initiated during an unsettled diplomatic relationship with France and were intended to combat internal threats against the U.S. government by protecting Americans from enemy powers—in this case, French aliens—but it served a secondary purpose in the suppression of criticism generally. The acts, however, faced growing opposition and sparked controversy as individuals were prosecuted under it. President Jefferson, who believed that the laws were unconstitutional on First Amendment grounds, later pardoned those who had been prosecuted (Smith 1956).

The hostilities of World War I prompted another surge of security mandates, leading Congress to enact the Espionage Act in 1917,[2] later amended to create the Sedition Act of 1918 and the Alien Control Act in 1918.[3] The growing fear of communist subversive and espionage activities in domestic affairs led to a broad application of the Espionage Act, and thousands of individuals were prosecuted. Although the Espionage Act of 1917 was meant to deal with the specifics of espionage and the protection of military secrets, some of the provisions were related directly to the suppression of freedom of speech. For example, section 3 of Title I of the act made it a crime for any person during a time of war to "make or convey false reports or false statements with the intent to interfere" with U.S. military success or "to promote the success of its enemies; cause or attempt to cause insubordination, disloyalty, mutiny, or refusal of duty, in the military or naval forces of the United States; or obstruct the recruiting or enlistment service of the United States." The Espionage Act was challenged in the context of *Masses Publishing Company* v. *Patten* (1917), when the postmaster general, with the authority granted to him under the Espionage Act of 1917, targeted the revolutionary journal *Masses* because the nature of the cartoons and texts it published were purportedly subversive in nature.[4] For instance, one of the poems that came under fire was a tribute to Emma Goldman and Alexander Berkman, both of whom were jailed for opposition to the war and the draft:

Emma Goldman and Alexander Berkman
Are in prison tonight,
But they have made themselves elemental forces,
Like the water that climbs down the rocks;
Like the wind in the leaves;
Like the gentle night that holds us;
They are working on our destinies;
They are forging the love of the nations; . . .
Tonight they lie in prison.[5]

When the journal requested an injunction from the trial, to be issued by the court to stop pending legislation, it was granted by Justice Learned Hand. Justice Hand aptly described the difficulty of striking a balance between security and the continued preservation of liberty, particularly freedom of speech: while "it may be that Congress may forbid the mails to any matter which tends to discourage the successful prosecution of the war,"[6] the claim of the postmaster, who said that the cartoons and poems tended "to arouse discontent and disaffection among the people with the prosecution of the war and with the draft tends to promote a mutinous and insubordinate temper among the troops," was too broadly construed. Hand went on to state that to read the word *cause* so "broadly would . . . involve necessarily as a consequence the suppression of all hostile criticism, and of all opinion except what encouraged and supported the existing policies."[7] The court of appeals later overturned Hand's injunction arguing that:

The question whether the publication contained matter intended willfully to obstruct the recruiting or enlistment service is less doubtful. Indeed, the court does not hesitate to say that, considering the natural and reasonable effect of the publication, it was intended willfully to obstruct recruiting; and even though we were not convinced that any such intent existed, and were in doubt concerning it, the case would be governed by the principle that the head of a department of the government in a doubtful case will not be overruled by the courts in a matter which involves his judgment and discretion, and which is within his jurisdiction.[8]

Unlike Hand, the appellate court argued that the possibility of obstructing recruiting efforts was sufficient to silence the journal and that the role of the court in this instance was not to second-guess the judgment of the department of the government. The Supreme Court later followed this logic in *Schenck* v. *United States* (1919), which upheld the conviction of the general secretary of the Socialist Party of Philadelphia for distributing a circular that urged men conscripted for military service to exercise their constitutional rights violated by the Conscription Act. In affirming the conviction, Justice Oliver Wendell Holmes argued that a nation at war creates a unique circumstance when civil liberties are necessarily restricted:

[T]he character of every act depends upon the circumstances in which it is done. The most stringent protection of free speech would not protect a man in falsely shouting fire in a theater and causing a panic. It does not even protect a man from an injunction against uttering words that may have all the effect of force. The question in every case is whether the words used are used in such circumstances and are of such a nature as to create a clear and present danger that they will bring about the substantive evils that Congress has a right to prevent. It is a question of proximity and degree. When a nation is at war many things that might be said in time of peace are such a hindrance to its effort that their utterance will not be endured so long as men fight and that no Court could regard them as protected by any constitutional right.[9]

The logic of enhanced security in times of war also found in the decision in *Abrams* v. *United States* (1919), in which the Sedition Act of 1918 was challenged.[10] In the majority opinion, Justice John Clarke stated that Abrams's actions interfered with the country's war efforts: "Even if their primary purpose and intent was to aid the cause of the Russian Revolution, the plan of action which they adopted necessarily involved, before it could be realized, defeat of the war program of the United States."[11] The Court described the purpose of the Sedition Act of 1917 as protection for the war program and the national security it was designed to achieve.

In 1919, the fear of an internal threat was fed by the revelation of a scheme to bomb notable individuals, including Supreme Court Justice Oliver Wendell Holmes and John D. Rockefeller. The next day, a series of riots occurred in Boston, New York, and Cleveland, among other cities, as police clashed with May Day celebrators. When, on June 2, 1919, eight bombs exploded simultaneously across the nation, including one at the house of Attorney General A. Mitchell Palmer, swift governmental reaction followed. In 1920, Palmer, with the assistance of the FBI's J. Edgar Hoover, ordered a widespread raid against the radical groups thought to be behind the attacks in eleven cities, including the Union of Russian Workers headquarters in New York. The Palmer raids took place against a political backdrop of a widespread fear of a Bolshevik revolution in the United States, and reports of the raids were met by an ecstatic public. The threat of a communist coup led to widespread societal acceptance of focusing the nation's investigative efforts on Russian aliens and their sympathizers (Hoyt 1969).

Public and political support waned as the realization of systematized overthrow of the government had largely been exaggerated (Stone 2004). After some two thousand prosecutions under the Espionage Act of 1917 and the Sedition Act of 1918, in which individuals were sentenced to anywhere from ten to twenty years in prison, the Sedition Act of 1918 was repealed by Congress (the Espionage Act of 1917 still stands). After the war, political support for continuing the prosecutions under the Sedition and Espionage acts began to dissipate, and in the 1930s Franklin Delano Roosevelt granted amnesty to all of those who had been convicted under the acts (Stone 2004), and the fear of communism began to diminish.

In 1934 President Roosevelt identified crime as a threat to internal security, pointing to an increase in bank robberies, kidnappings, and extortions. In response, Congress enacted the Twelve Point Crime Program proposed by Attorney General Homer Cummings, including federal legislation to address bank robberies, racketeering, murder or assault of federal officials, and extortion. The increase in federal powers to deal with crime was widely accepted by a public that had been convinced by the images in media and film depicting high-profile kidnappings and bank robberies that crime was a pandemic. The notorious John Dillinger became a symbol for the Twelve Point Crime Program, and his eventual dramatic murder provided a triumphant justification for the increase of federal power (Theoharis 2004).

The cold war reintroduced the specter of an internal threat that championed security objectives. On June 28, 1940, Congress passed the Alien Registration Act of 1940, commonly known as the Smith Act, which made it "unlawful . . . to print, publish, edit, issue, circulate, sell, distribute, or publicly display any written or printed matter advocating, advising, or teaching the duty, necessity, desirability, or propriety of overthrowing or destroying any government in the United States by force or violence."[12] Closely connected with the objectives of the Smith Act was the Voorhis Act of 1940,[13] which required registration of the members of organizations with the attorney general According to the Voorhis Act, an organization in the United States must register if it's organization's goal was "the establishment, control, conduct, seizure, or overthrow of a government or subdivision thereof by the use of force, violence, military measures, or threats of any one or more of the foregoing." Significantly, registration included the addresses of the organization's chapters and the names of every individual "who has contributed any money, dues, property, or other thing of value to the organization or to any branch, chapter, or affiliate of the organization."

The McCarran Internal Security Act of 1950, which followed, was directly aimed at unearthing communists and their sympathizers. The act was passed over the veto of President Truman, who contended that it "make(s) a mockery of the Bill of Rights and of our claims to stand for freedom in the world" (Veto of the Internal Security Bill, 1950).

Communist organizations were not the only targets. The National Association for the Advancement of Colored People, the National Maritime Union, and the United Automobile Workers Union also were put under surveillance (Theoharis 2004). The FBI indexed newspaper subscribers and distributors and recipients of socialist literature, and it identified signatures on resolutions or protest petitions. More invasive search techniques, such as wiretaps and undercover informants, were employed to monitor known socialist activists (Schmidt 2000). In addition, there was a belief that the motion picture industry, because of its influence on public opinion, should be put under investigation. J. Edgar Hoover intensified this focus on the motion picture industry when *For Whom the Bell Tolls* and *Hangmen Also Die* were released in 1943, perceiving them as indicative of subversive sentiment (Theoharis 2004).

The concerns of the red scare of the interwar period lingered in American politics, only to peak again with the proceedings of the actions of the House Un-American Activities Committee from 1947 to 1954. Executive Order 9835, creating federal loyalty review boards, set McCarthyism on its course. Counterintelligence tactics were also directed at the emergence of black nationalist groups in the upsurge of political resistance in the civil rights era. The counterintelligence program COINTELPRO was used from 1956 to 1971 to "expose, disrupt, misdirect, or otherwise neutralize the activities of black nationalist, hate-type organizations and groupings, their leadership, spokesmen, membership, supporters, and to counter their propensity for violence and civil disorder" (Theorharis 2004, 122).

Even before its linkage to communist activity, the FBI was attempting to control and manage the civil rights movement. Although this linkage was most visible in the years of the Kennedy administration, counterintelligence activity was focused on subversive activity connected to race as early as FDR's second term. In 1939, for instance, Roosevelt empowered the FBI to investigate a broad array of civil rights cases, ranging from voter fraud to police brutality (O'Reilly 1988). The same type of approach was directed to political resistance to the Vietnam War in the belief that political resistance provided feeding ground for the communist movement.

Change and Continuity after September 11

These historical antecedents are important to a consideration of points of change and continuity in the post–September 11 environment. Did, as some argue, Americans lost their ideological moorings after September 11? According to Rosen (2005), for example, a crowd mentality leads individuals to accept an architecture of surveillance without rationally thinking about the detrimental effect on privacy and civil liberties: "We are living in the world of graphomania; we are experiencing the constant din of intimate typing—in e-mail, in chat rooms, in blogs, and in the workplace . . . the crowd wants what it wants, and one thing it wants is personal exposure from anyone on whom it fixes its irresistible gaze" (191). Does this pared-down explanation of the rise of the information technology coupled with the events of September 11 capture the complexity of the current era?

Attitudes toward and perceptions of biometric technology can shed light on the question. As in many prior times in U.S. history, political support of greater security gave way to the questioning of it. As a tool of security, biometric technology became a touchstone for the many public policy debates involving concerns of individual liberty, increased presence of governmental surveillance, legitimate paternalistic intervention in the name of harm, and trust and confidence in institutions. These long-standing concerns did not simply fall away in the post–September 11 environment as some would have us believe.

Biometrics: Political Promise and Peril

It might appear at first glance that the immediacy of the terrorist threat made real by the attacks on the World Trade Center and the Pentagon, and the downed airplane in Pennsylvania, would have given the government carte blanche to use biometric technology for the purposes of identity verification. Indeed, there was a loud call from the American public after September 11 to deal with the threat of terrorism. The answer came in the form of legislation mandating, among other things, a system of identification that was failsafe. Specific recommendations for the inclusion of biometric technology as a tool of security were written into both the USA PATRIOT Act of 2001 (P.L. 107-56) and the Enhanced Border Security and Visa Entry Reform Act of 2002 (P.L. 107-173), which together set the stage for a debate about the role of biometric technology in the war on terror. But this debate was about much more than the technology, as the discussions soon revealed.

In response to the legislative mandates, the government started to move ahead. Senator Dianne Feinstein (D, California) led the call to action in a hearing of the Subcommittee on Technology, Terrorism and Government Information, held November 14, 2001, the title of which was revealing: "Biometric Identifiers and the Modern Face of Terror: New Technologies in the Global War on Terrorism." Feinstein explained the potential of biometric technology in the war on terror at the outset of the hearing, unveiling the high expectations:

After the September 11 attacks many Americans began to wonder how the hijackers were able to succeed in their plans. How could a large group of coordinated terrorists operate for more than a year in the United States without being detected and then get on four different airliners in a single morning without being stopped? The answer to this question is that we could not identify them. We did not know they were here. Only if we can identify terrorists planning attacks on the United States do we have a chance of stopping them. And the biometrics technology, the state-of-the art technology of today, really offers us a very new way to identify potential terrorists. (Hearing 2001)

The potential for biometric technology was far reaching in many sectors. Biometric technology, because of the use of aircraft in terrorist attacks, naturally was advocated for as part of airline security. As Monte Belger, acting deputy administrator of the Federal Aviation Administration explained,

The bottom line from the FAA is that biometric technology has the potential to improve aviation security and these systems are eligible for funding under the airport improvement program. As we move ahead, I think we should keep in mind that there probably is no one solution, that probably technology by itself will not be the solution to the issues that we are facing, but these technologies hold great promise. . . . There are some significant challenges and in the world of aviation security we are anxious and willing and want to get involved to address these challenges and make these systems become operational at our nation's airports. Our fundamental goal is 100 percent screening of all passengers, baggage, airport and airline personnel, and we believe that these systems have a role in the future. (Hearing 2001)

The possibility of using biometric identifiers for airline safety was not the only considered deployment. Joanna Lau, of Lau Technologies, testified about the potential for biometric technology in border security as a defense against terrorism. She described the crisis of September 11 as an urgent situation akin to Desert Storm in the First Gulf War:

I want to say that it is unfortunate that it often takes a crisis to create an opportunity to make change. About 11 years ago my company was involved with Desert Storm and that certainly brought us a tremendous opportunity during that crisis and also gave us the opportunity to learn about the defense and learn more about technologies, how it could improve our nation. Well, we are now here at an urgency to really make change because even though we win the war, we are going to create more terrorists around us, so it is important we make change at our borders, as well as what is going on here—not only terrorism, also the most wanted list that could be endangering us domestically, as well. (Hearing 2001)

The situational use of biometric technology in airports and at borders was translated into a broader goal of protecting society as a whole. Visionics, an American manufacturer of face recognition technology, in a white paper, "Protecting Civilization from the Faces of Terror," touted the benefits of facial recognition technology and framed the obstacles to its success in a large-scale deployment as one of federal funding. In fact, the white paper claimed that facial recognition technology was ready to "immediately spot terrorists and prevent their actions" (Visionics 2001). In the words of one member of biometrics community, "September 11 created a long-awaited moment for the biometric industry" (Feder 2001). In fact, the potential size of the biometric market was estimated to exceed $4 billion in the United States in 2007, representing an 80 percent growth in the market (Ciarracca 2002).

Leaders in the biometric industry were now participants in the debate regarding the policy steps to be taken in the aftermath of September 11. Although the International Biometrics Industry Association had been established in 1998 and the Biometric Consortium in 1995, industry leaders suddenly found themselves in the limelight as the government searched for a solution to undermine terrorist activities. The Biometrics Consortium and the Biometrics Foundation, two central organizations in the biometric community, emphasized the need to promote security and positioned biometric technology as a tool. In fact, the Biometrics Foundation argued that the United States was ahead of other countries in developing the technology but behind in terms of implementing it. It advocated widespread use of biometric technology in "passports, visas, identification cards, and other travel documents" (Collier 2001). This was music to the ears of members of Congress and the Bush administration, who were hurriedly searching for a solution to a growing recognition of potential security risks.

Optimistic rhetoric was used to describe the possibilities proffered by biometric technology. The technology was described as an unobtrusive means to assist the war on terror because it could provide protection in a wide variety of contexts. It was thus poised to promote security by verifying identities at ports of entry, critical infrastructures, airports, governmental buildings, and a whole host of private applications.

The biometric community was very much aware of the potential market created by September 11 but also aware of the obstacles that had to be overcome. One of these was the difficulty of winning social acceptance. Security and safety were one set of concerns, but regulatory compliance with privacy rights and the convenience of consumers were also significant concerns. In a statement prepared by Identix, the difficulty and necessity of striking a balance was obvious:

> Providing protection against these threats presents a special challenge. Because airports support activities that are both public—passengers, visitors and airport employees—and private—such as air cargo and mail—these locations are part transportation hub, shopping mall and industrial complex. As a consequence, requirements for public safety and security are a hybrid of both commercial and industrial needs similar to a small or medium-sized U.S. town or city. The resulting challenge is to balance security, safety and government regulatory compliance with the privacy rights and convenience of individuals. (Hearing 2001)

The issues of safety, security, privacy, and convenience were made more complicated by the very nature of the biometrics industry. The fragmentation of the industry and the consequences of this disorganization on the possibility of using biometric technology on a large scale were of great concern to the Feinstein subcommittee. Senator Feinstein commented:

> I have received a number of phone calls from experts on biometrics and these experts, including the main biometric industry associations and the National Security Agency, have suggested that the industry is extremely fragmented, lacks minimal standards, and does not work well together, given the hypercompetitiveness of the companies. Currently, for example, there are about 140 companies trying to sell hundreds of different overlapping biometric devices of multiple types. These include fingerprints, hands, irises, faces, retinas, voice, handwriting, et cetera. (Hearing 2001)

As quickly became clear, biometric technology encompasses myriad technologies, each with its own set of weaknesses. Instead of one coherent technology, the biometric industry was a series of industries within an industry. This meant that the process of achieving stability and reliability of the technology was more complicated than it might have been if there had been only one technology associated with biometrics. Questions of reliability, accuracy, and compatibility were multiplied across the various sectors of biometric technology. Certainly the lack of stability and coherence in the industry might have been endemic to any technology reaching technological maturity; however, these factors added to the issue for decision makers in the aftermath of September 11.

This complicated mix of vendors meant that the industry had not reached any internal agreement about the best biometric technology and had not agreed on any evaluative standards by which to make this judgment. And the deployment of one biometric technology would not provide the ultimate solution. As Joanna Lau explained:

> True, there are a lot of technologies in place but not one single technology is going to provide your silver bullet. And the other thing is that every application and environment is very different and we rely on some of the experts that you have working in your government to work with industry. We are here to offer our expertise and help but we are not taking over their job. We have to work with them. (Hearing 2001)

The discussions that followed made it clear that each biometric technology had disadvantages and advantages when compared to the others. Within the segmented biometric industry, many claimed to have the best technology. Each sector of the industry had statistical evidence demonstrating the success rate of its respective biometric system. The problem was that these small-scale deployments could not be translated easily to large-scale deployments. A seemingly small error rate would quickly become significant if biometric technology was deployed on a large scale. Also complicating the large-scale deployment of biometric technology was the fact that the biometrics industry would require extensive cooperation with the U.S. government in terms of testing, standard setting, deployment, and cooperation in information sharing. Accordingly, from the perspective of Feinstein's committee, one potential solution seemed to rest in the establishment of a central testing facility that would combine the resources of academia, government, and industry to provide some coherence to the biometrics industry. She proposed that such a center

> would involve the leading private sector biometric institutions. These would include the International Biometric Industry Association, the Biometric Foundation, the Center for Identification Technology Research at West Virginia University, the leading university biometrics center. The National Security Agency would be the initial coordinating agency for this center and could, at the president's discretion, be replaced by the Office of Homeland Security. (Hearing 2001)

Senator Feinstein attracted attention when she characterized the establishment of a federally chartered center for the testing of biometric technology as similar to the governmental role in the Manhattan Project (Hearing 2001). These analogies were disturbing to those who were already concerned about a burgeoning federal government empowered with legislation that mandated the wide-scale use of biometric information.

Not surprisingly, the need to make use of existing databases, coupled with necessary governmental involvement, pointed to the expansion of the FBI's Integrated Automated Fingerprint Identification System (IAFIS) to include visa applicants and visa holders who were wanted in connection with a criminal investigation. The prospects were promising in the opinion of Michael Kirkpatrick, assistant director in charge of the FBI's Criminal Justice Information Services Division. Testifying at the hearing, Kirkpatrick explained:

> Since the IAFIS is the world's largest biometric database with an infrastructure which already connects local, state and federal agencies, it is a tool that could be used to move our country's security perimeter beyond our borders. While the FBI believes that the IAFIS is a national asset, its development has also had significant international ramifications. On a global front, fingerprints are the most widely held and used forms of positive identification. In this regard the FBI has taken the lead in an effort to develop international standards for the electronic exchange of fingerprints. We frequently meet with our colleagues in the Royal Canadian Mounted Police and United Kingdom, as well as in Interpol, on this topic. Technology for the capture, search, storage and transmission of fingerprints is widely available and, as you will hear today, becoming more economical. Fingerprint databases already exist at the federal, state and local levels and all existing criminal history records are based on fingerprints. (Hearing 2001)

The existence of criminal and civil records of fingerprints presented a ready-made test bed and an existing database from which comparisons could be drawn. From the criminal files, fingerprints existed as a result of arrests at the city, county, state, and federal levels. From civil files, fingerprints are acquired related to background checks for employment, licensing, and other noncriminal justice purposes as authorized by federal and state law and in compliance with appropriate regulations. Because of the time constraints imposed by the USA PATRIOT Act and the Enhanced Border Security Act, the IAFIS system made sense logistically, but its use heightened the concerns of the development of a "big brother" system.

These fears were also exacerbated as "failures" and public resistance to biometric identification systems began to arise. One notable use of facial recognition system was Superbowl XXXV in Tampa in January 2001. Critics were quick to point out that the system was not reliable in picking out "terrorists" and that surveillance at public events amounted to a serious violation of Americans' civil rights. This criticism was not without cause. Facial recognition is challenged in open environments where the capture of the image is affected by the surrounding conditions. The lack of quality in the image in turn affects the ability of the system to screen potential terrorists, which was, perhaps unrealistically, the expectation of the public. Reports of fallibilities began to erode the high expectations that the biometrics industry had for the market for biometric technology and tempered the rhetoric of success that had predominated in the immediate aftermath of September 11. Critics were aligned against biometric identification systems because of the association with expanding governmental presence. Generally the public debate about the consequences of the increasing presence of information technologies in our lives was growing. As Daniel Solove (2001) explained, "We are in the midst of an information revolution, and we are only beginning to understand its implications. The past few decades have witnessed a dramatic transformation in the way we shop, bank, and go about our daily business—changes that have resulted in an unprecedented proliferation of records and data" (1).

While September 11 clearly raised the prominence and familiarity of biometric technology and spurred the advance of technological maturity, questions remained about societal acceptance. Ironically, the notoriety of biometric technology made it more prone to high expectations and susceptible to a range of criticisms. These criticisms were directed not only toward the technology. The potential upside of the technology and the costs of failure, both societal and political, were extremely high given the call for greater security after September 11. Errors in deployment would have profound consequences on individual privacy and civil liberties and trust and confidence in the institutions that implemented biometric technology. Critics were to ready to pounce on the failures and consequences of surveillance technologies in the hands of the newly created Department of Homeland Security. The expanding bureaucratic presence of the department and the technological tools at its disposal also gave rise to an evaluation of the appropriate level of paternalistic presence in our lives.

Societal Perceptions and the War on Terror

Privacy advocates latched onto the Orwellian rhetoric of "big brother" and began attacking biometric technology as the actualization of the

Benthamite Panopticon. Drawn from the concept introduced by the philosopher Jeremy Bentham, and later popularized by Michel Foucault, biometric technology was viewed as an architecture of surveillance. Bentham had proposed a prison architecture based on a simple idea: implied surveillance, which translated into the control of the prisoners. All inmates could see the centrally located tower, and the windows of the tower were positioned so that if someone were inside they could see into every cell, yet the inmates could never determine with certainty whether they were being observed by anyone in the tower, creating the conditions of constant surveillance with architecture and evoking self-discipline among the inmates

For many of its critics, biometric technology functioned as the Panopticon. The perception that biometric technology had the potential to be all-seeing would create the desired effect of discipline because individuals would never know with certainty whether they were being surveilled. To the critics, the result would be a totalitarian society where thoughts and behavior would be censored by the constant fear of surveillance.

These concerns were not entirely new. Before September 11, privacy advocates had already articulated a well-developed critique of the lack of regulation for the protection of liberty interests in light of the growing information flow in both the public and private sectors. The sectoral protections offered by the Health Insurance Portability and Accountability Act (HIPAA) and Gramm-Leach-Bliley[14] were thought to be only temporary solutions for a larger problem of information acquisition and sharing. Concern over individuals' sharing large amounts of information with private credit card companies, cell phone companies, and financial institutions now became an objective of surveillance, encroachment of governmental presence on civil liberties, and a lack of trust and confidence in institutions to guard personal information in light of the imperative of terrorism. The White House announced the need for information sharing to fight against terrorism on its Web site:[15]

> As the terrorist attacks on transportation infrastructure in London and Madrid demonstrate, critical infrastructure can be a prime target for the transnational terrorist enemy we face today. The private sector owns and operates an estimated 85% of infrastructure and resources that are critical to our Nation's physical and economic security. It is, therefore, vital to ensure we develop effective and efficient information sharing partnerships with private sector entities. Important sectors of

private industry have made significant investments in mechanisms and methodologies to evaluate, assess, and exchange information across regional, market, and security-related communities of interest. This Strategy builds on these efforts to adopt an effective framework that ensures a two-way flow of timely and actionable security information between public and private partners.

This criticism extended easily to biometric identification systems that could amass and link massive amounts of personal information, possess the potential to be used covertly, and promised the capability of surveillance. The interest in biometric technology, evidenced in legislation that pointed to the future of the use of biometric technology in the management of borders and driver's licenses, and to monitor airline passengers, coupled with the rhetoric that promoters began to attach to the technology in order to align it with the public support for the war on terror, made it an easy target.

Organizations such as the American Civil Liberties Union (ACLU), the Electronic Frontier Foundation, and the Electronic Privacy Information Council were poised to object to the proposed uses of biometric technology. The privacy and civil liberty concerns of the increasing presence of information technology in our lives fit neatly with the fears associated with biometric technology. And the political divide that resulted in the aftermath of September 11 between those who favored the war on terror and those who feared the increased powers of the U.S. government served only to stigmatize biometric technology. Some of the criticisms were fair, others overstated. The Electronic Frontier Foundation described the potential misuse of biometric technology as far reaching. The organization of biometric identifiers into databases that could potentially be used to track the behavior of individuals would lead to the chilling of democratic participation, an increase in the "circumstantial evidence" available for criminal prosecution, and the correlation of the behavior of individuals against predetermined patterns that could be used to manipulate behavior and ultimately track individuals, to the detriment of a free democratic society (Electronic Frontier Foundation 2001).

In a letter to Senator Ernest Hollings on the issue of biometric identifiers, the ACLU advocated the deletion of Section 302 of the Aviation Security Improvement Act, which would allow the Transportation Security Administration (TSA) to establish "identification verification technologies," including facial recognition and other biometric identification technologies, for the screening of airline passengers (ACLU 2001). The ACLU framed the introduction of biometric identifiers in terms of unfettered surveillance and information gathering, arguing that "we enter the brave new world without any rules. There are no legal controls over how biometrics can be used—whether the information can be sold, whether it can be turned over to law enforcement without a warrant" (ACLU 2001). In addition to these concerns, the function creep of biometric identifiers continued to be a pressing problem. Barry Steinhardt, associate director of the ACLU, asked, "Can there be a secondary use? What is the form of consent, and is it truly voluntary? What security is there against theft? The paramount problem is that the technology has been developing at light-speed, while the law has developed not at all" (ACLU 2001).

The concerns regarding the integration of biometric identifiers as a security measure were tied to myriad fears, one of which was the eventual institution of a national identification card. From the perspective of the ACLU, "This type of nationwide biometric identification system would not increase security and would pose a serious threat to privacy and civil liberties." The ACLU went on to make the connection between the possible integration of biometric identifiers and September 11: "One reaction to the terrible events of September 11 was renewed discussion about instituting a national ID as a counter-terrorism measure. Although national ID proposals received fierce debate in the fall, the Administration and Congress wisely rejected them. Direct passage of a national ID card, however, is only one possible path to such a system" (ACLU 2001). Whether these concerns aptly portrayed the realities of biometric technology was of little consequence as the fears surrounding its potential escalated.

But for September 11, Would We Be Willing to Accept Biometrics?

Absent the context of September 11 and its effects on the existing concern regarding the potentially invasive presence of information technologies, biometric technology might have made a gradual, and perhaps less controversial, entrance into public policy debates. In fact, biometric technology had already been put to use in several pieces of legislation long before September 11. The Truck and Bus Safety and Regulatory Reform Act of 1988 required "minimum uniform standards for the biometric identification of commercial drivers."

Both the Personal Responsibility and Work Opportunity Act of 1995, a welfare reform law, and the Immigration Control and Financial Responsibility Act of 1996, an immigration reform law, called for the use of "technology" for identification purposes. In addition, the Illegal Immigration Reform and Immigrant Responsibility Act of 1996 required the U.S. Immigration and Naturalization Service to include on alien border crossing cards "a biometric identifier (such as the fingerprint or handprint of the alien) that is machine readable."

These one-to-one matching systems, which serve to identify participants enrolled in a biometric identification system, were in fact successfully implemented in a variety of settings and in increasing numbers. For example, the use of biometric identifiers as a way to authenticate individuals in financial transactions, welfare distributions, or time and attendance verification or to monitor physical access were well underway prior to September 11 and generated little controversy. The sudden prominence of biometric technology after September 11 highlighted the possibilities of the technology and its potential failings in one-to-many matching situations. But did this matter to the American public? Many have argued that September 11 provided a window of opportunity for the societal acceptance of biometric technology, leading the public to set aside concerns of technological surveillance in favor of achieving security. Jeffrey Rosen (2005) contends, for instance, that "before 9/11, the idea that Americans would voluntarily agree to live their lives under the gaze of a network of biometric surveillance cameras, peering at them in government buildings, shopping malls, subways, and stadiums, would have seemed unthinkable, a dystopian fantasy of a society that had surrendered privacy and anonymity" (34). Did September 11 translate into Americans' surrender of individual autonomy to the tools of technological surveillance?

For the United States, the internal threats to social stability, in the form of criminality or immigration, served to substantiate the need for a system of identification. But it is not only criminal threats to social order that justify identification. Another basis for instituting a system of identification can occur in the face of external threats to society, as in the case of war.

In the British case, the advent of a world war provided a ready purpose for identification. "A war that demanded mobilization of nearly all industry and population inevitably led to an expansion of the state in Britain:

political actions that would have been unthinkable before 1914 now became necessities" (Agar 1991, 103). In light of the impending need for military enlistment, the National Registration Act of 1915 was driven by the bureaucratic need to gauge the potential number of available men for industrial and military purposes. The act was, from the perspective of the state, a prelude to conscription. The Military Service Act followed in 1916 when it became clear that volunteerism would not meet the needs of the war. Although the Military Service Act of 1916 eventually delivered potential enlistees, it was the extension of the National Registration Act of 1915 for purposes other than war that caused a lack of cooperation. With the need of enlistees diminished, the administrative purpose promoted in the second introduction of the National Registry was food. Described by Registrar-General Sylvanus Vivian in this way:

> Any system of NR, as being an instrument of conscription, would obviously be received by the public with some reserve and suspicion, and in its actual administrative working, when established, would be exposed to a hostile bias on the part of the individual members of the public. By linking that system with the equally necessary system of registration for food rationing purposes . . . motives would be interlocked. . . . In the case of machinery serving both purposes these risks would cancel out and the interlocking motives would make for far greater effectiveness in both respects. (Agar 1991, 108)

The justification of ending food shortages and potentially providing a means to maintain contact with evacuees did not win public approval. While it existed, the national registration card served other administrative purposes, becoming almost essential for verifying identity in other sectors of life. By the end of the war, however, the rhetoric of food shortages began to lose its salience. Although an attempt was made to redirect the rhetoric of national identification for health and welfare under the auspices of the National Health Service, a legal challenge ultimately felled the system of registration (Agar 1991). When the validity of the National Registration Act was challenged in peacetime, public sentiment turned against the identification system. Although the King's Bench Divisional Court rejected the appeal of the decision of the magistrates who had upheld the National Registration Act, the idea of the identity card had become an unacceptable exercise of state power within a relatively short period of time (Agar 1991). Vivian (1915) described the likely cause of its demise as a loss of freedom and privacy on the behalf of individuals subject to the registration:

> This is not due to any lack of patriotism or of respect for the law, but has its cause deep down in the genius of the nation, the freedom of the private life from bureaucratic intrusions, its unfamiliarity with and distaste for formalities or procedure and "red tape." Such a system could only be successful when enforced, as in Germany, by a rigorous and ubiquitous police system upon a nation accustomed to be regulated in all minor matters of life. Any system of registration which is intended to operate successfully in this country must be based upon different principles. (Agar 1991, 109)

Several important points arise out of the British effort. The nature of modern identification requires that the state articulate an objective—but not any objective will do. The rhetoric of war and the need for enlistees, for instance, were persuasive to the citizenry and a system of identification obviously necessary for the state to remedy the situation. In the aftermath of war, when the rhetoric of justification was shifted toward general health and public welfare, the need for identification did not seem as legitimate or pressing to the citizenry, and public support quickly faded. Yet despite citizen resistance to the continuation of the identification system, the potential benefits of identification were many for modern bureaucracy. The ability to maintain a statistical picture of the nation with national registration would facilitate coordination and cross-reference of smaller registers, which would enable government departments to work together easily (Agar 1991). The benefits to the state, and ostensibly for its citizenry, were set aside when the public was not willing to see it as beneficial to the relative cost of potential state intrusion. The benefits of a system of identification and the paternalistic intervention that they represent must be able to sustain public support. In the case of Britain, paternalistic intervention was acceptable during crisis. Once this crisis dissipated, public support quickly waned too.

The vagaries of public support can be explained in a variety of ways. In the first instance, the legitimacy of paternalistic intervention on behalf of the citizenry because of an external threat was an expected function of the state, and there was a diminution of concern related to a loss of freedom because freedom itself was at risk. As the external threat diminished, public support for the program dissipated. This was also the reaction to a proposal of universal fingerprinting in the United States. Critics like the ACLU were quick to denounce J. Edgar Hoover's program as a violation of civil rights despite his efforts to garner social support with tales of criminal activity and the prospects of a revolutionary takeover by communists. Internal security, whether used as a justification to prevent criminal behavior or maintain social order, is a strong incentive for public support. This rationalization is ripe for public support, in part, because the class of individuals subject to the system of identification is limited and, in eyes of many, deserving of scrutiny.

A system of biometric identification is likely subject to the same societal limitations imposed on those that preceded it. The rhetoric of the war on terror, for example, serves as a source of legitimacy for using biometric technology, but it does not serve to justify all instances of its use. There has been support for the use of biometric identification systems at border control, but its use in public events like the Superbowl has not gained the same public support. As in the case of public acceptance of fingerprinting criminals in U.S. history, there is also support for identification of certain classes of individuals and of terrorists, even if it might involve a system of biometric identification with screening that may make some misidentifications. Where support tends to wane is not unlike its historical referents. If the system of identification is generalized and wide scale, involving information gathering on activities, even if carried out in public, then the legitimacy and the benefit to be gained with identification are lacking relative to the perceived loss of freedom.

September 11 and the American Public

The events of September 11 have added fuel to the argument that the United States needs a new system of national identification, including the use of biometric identifiers. Certainly September 11 was a profound event for Americans that brought an end to the sense of invulnerability that Americans largely shared following the end of the cold war. Yet to understand the dimensions of societal acceptance of biometric technology and the effects of September 11 requires a better understanding of the responses evoked by biometric technology. We must examine them for our insights not only about the technology but also for what they can reveal about the United States. Biometric technology, given its ascendance in the post–September 11 environment and the information revolution, is telling of much more than a static assessment of societal perceptions. Before we announce that the revolution has taken place because of either September 11 or the information age, let us remember that this longing to see points of change as revolutionary is endemic to the human condition. As Bernard Yack (1992) so aptly put it:

> There are moments in our lives when we cannot help but be filled with disgust at the dangerous, vulgar, and unruly aspects of human society. At these moments, we long to escape the limitations of our social existence. Since such sentiments are probably coeval with civilized life, it is not surprising to discover that there has been a longing for social transformation in every culture and every historical epoch. On the other hand, the intensity, direction, and impact of this longing vary enormously from one epoch and culture to another. This longing for social transformation can manifest itself as a gentle nostalgia for a distant past or as a violent yearning for the Apocalypse; it can focus on particular objects and institutions or make us uneasy with all of our social relationships; it can provide an individual with a pleasant diversion or inspire mass movements (3)

The need to see revolutionary changes obscures the nature of the evolution. Although biometric systems of identification are a relatively new technology and the struggle against terrorism and extremism is enduring, there are historical analogies that illustrate the ongoing tension between liberty and security that the current debate is part of and from which insight can be drawn.

3 Privacy and Biometric Technology

As every man goes through life he fills in a number of forms for the record, each containing a number of questions. . . . There are thus hundreds of little threads radiating from every man, millions of threads in all. If these threads were suddenly to become visible, the whole sky would look like a spider's web, and if they materialized as rubber bands, buses, trams and even people would all lose the ability to move, and the wind would be unable to carry torn-up newspapers or autumn leaves along the streets of the city. They are not visible, they are not material, but every man is constantly aware of their existence. . . . Each man, permanently aware of his own invisible threads, naturally develops a respect for the people who manipulate the threads.

—Alexander Solzhenitsyn, *Cancer Ward* (1968)

Privacy concerns are inseparable from biometric technology, but how biometric technology affects and is affected by privacy is complicated. Privacy is a multifaceted concept that finds its way into discussions about private property, intimate decision making, secrecy, personhood, and the limits of governmental intervention. Add to this complexity the wide variety of settings in which biometric technology can be deployed, and a vast array of privacy concerns is triggered. Societal assessment of biometric systems of identification hinges not only on the diverse meanings of privacy that individuals pin on their interests, but on the many different points of interaction they might have with biometric technology in a diverse set of public and private institutional settings.

The interaction between biometric identification systems and the many interests associated with privacy requires an understanding of how people respond to the technology and a consideration of the values of privacy that inform their perceptions. Rather than trying to build an abstract theory of privacy absent a consideration of context, the approach taken here is pragmatic, focusing on the experience of individuals instead of positing a universal theory. The intention is to understand the complexity of the relationship between privacy and biometric identifiers from the perspective of individuals, even if their understanding of biometrics is less sophisticated than that of a technologist and even if their comprehension of privacy does not align with the position of privacy advocates. It is at the level of societal perceptions that the nuances of privacy can be explored because it is around risks, challenges, and problems that the societal understanding of privacy congeals. Solove (2009) described privacy as a

> set of protections from a plurality of problems that all resemble one another, yet not in the same way. The value of protecting against a particular problem emerges from the activities that are implicated. . . . The value of privacy is not absolute; privacy should be balanced against conflicting interests that also have significant value. The value of privacy is not uniform, but varies depending upon the nature of the problems being protected against. (100)

New technologies, increasing reliance on personal information, and the political demands of security and liberty, together and in opposition, have prompted debate about the place of privacy in our lives. But how do we best understand the implications of these changes for societal perceptions of privacy and the technological or policy guarantees it demands? While biometric technology might evoke privacy concerns, these same societal impressions of biometric technology can be used to understand the points of obduracy and transformation in the meaning of privacy driven by changes, whether they are social, cultural, political, or technological. Understanding the apprehensions and hopes associated with a technology can serve as a window into norms, values, and expectations and, more important, guide the development of prudent public policy. The same understanding can serve as a tool of technologists for the development of socially acceptable forms of technology.

The use of biometric technology as a device to comprehend the social imperatives of privacy is informed by a notion of the social described by Bruno Latour (2005): "The 'social' designates two entirely different phenomena: it's at once a substance, a kind of stuff, and also a movement between non-social elements" (159). To study the social aspects of biometric technology is not to approach the inquiry with the expectation that the social elements are static and unmoving; it is to argue that the social elements of privacy are visible within discussions of biometric technology as it is evaluated, assessed, described, and instituted by society, revealing the social in the face of controversy and destabilization. In this sense, the role of biometric technology after September 11 and in the context of the information age presents a point of destabilization and controversy against which the meaning of normative privacy concerns can be better understood and dealt with in formulations of policy directed toward biometric technology. As Latour explained, "The 'easy' social is the one already bundled together, while the 'difficult' social is the new one that has yet to appear in stitching together elements that don't pertain to the usual repertoire" (165).

Understanding privacy in relationship to biometric technology requires tracing the social values of privacy not already bundled together in the law. There is more to privacy than its legal guarantees. Societal values of privacy are not always reflected in the current state of legal doctrine, as changes in the landscape of technology, science, or political mores are codified into law in fits and starts. The text of the legal doctrine of privacy is only a point of departure for comprehending the values of privacy, because the very nature of the language of law makes it difficult to capture social meaning as it inevitably changes. This difficulty of understanding the changing nature of the social perceptions of privacy is not a problem particular to biometric technology; the disjuncture between text and meaning is endemic to language.

The Norms, Values, and Expectations of Privacy

Legal guarantees of privacy are a starting point for understanding privacy, but the text of the law does not fully represent the normative aspects of privacy because of the long-recognized difficulty of indeterminacy in the text of the law. The critique of a strict notion of legal formalism was most persuasively made by H.L.A. Hart in *The Concept of Law* first published in 1961. Hart, a legal positivist in his own right, departed from the strict notion of legal formalism depicted by John Austin, who attempted to distill the law into legal rules. The semantic theory of Austin, a nineteenth-century lawyer, made following the law a necessarily dogmatic exercise: "The existence of law is one thing; its merit or demerit is another. Whether it be or be not is one enquiry; whether it be or be not conformable to an assumed standard, is a different enquiry. A law, which actually exists, is a law, though we happen to dislike it, or though it vary from the text, by which we regulate our approbation and disapprobation" (Austin 1832, 157).

Hart (1983) criticized Austin's idea of legal positivism, which conceived of legal rules as fixed and unchanging, and instead advocated a hermeneutic approach to understanding the application of the law:

The fundamental error [of legal formalism] consists in the belief that legal concepts are fixed or closed in the sense that it is possible to define them exhaustively in terms of a set of necessary and sufficient conditions; so that for any real or imaginary case it is possible to say with certainty whether it falls under the concept or does not; the concept either applies or it does not; it is logically closed (begrenzt). This would mean that the application of a concept to a given case is a simple logical operation conceived as a kind of unfolding of what is already there, and, in simpler Anglo-American formulation, it leads to the belief that the meaning of all legal rules is fixed and predetermined before any concrete questions of their application arises. (265)

The error Hart identifiedcan be extended to the relationship between biometric systems of identification and the legal concept of privacy. If the textual guarantees of privacy are applied rigidly to biometric technology, the fit may not account for the societal perception, or normative distillation, of the idea of privacy. This is because all legal concepts are wrapped in indeterminate meaning that are grounded in societal perceptions and differences of factual composition. Hart points to our inability, as a function of language generally, to frame rules of terms and expressions without avoiding the pitfalls of indeterminacy. Recognizing the inability to see all possible combinations of future factual or normative interpretations of a legal concept should serve as a warning against our exclusive reliance on legal doctrine to formulate answers to new factual applications. Hart (1983) claimed that we are incapable of defining concepts so exhaustively as to cover all imaginable possibilities. He cautions in his 1984 work against a simple application of legal doctrine that blindly prejudges legal issues whose factual composition and social consequences cannot be known.

Fitting privacy to the many existing and potential applications of biometric technology thus requires a discussion of normative expectations that are subject to changing conditions. Ludwig Wittgenstein (1978) suggested that it is essential to comprehend the "form of life" represented in the practices and normative understandings of it:

> When a man who knows the game watches a game of chess, the experience he has when a move is made usually differs from that of someone else watching without understanding the game. (It differs too from that of a man who doesn't even know that it's a game.) We can also say that it's the knowledge of the rules of chess which makes the difference between the two spectators, and so too it's the knowledge of the rules which makes the first spectator have the particular experience he has. (49)

As Wittgenstein (1968) urges, "Ask yourself: on what occasion, for what purpose do we say this? What kinds of actions accompany these words? In what scenes will they be used: and what for?" (para.23). Understanding privacy, like the game of chess, is not a matter of grasping some inner essence of meaning embedded in the text of the law; rather, privacy is defined by how and when individuals think it does or should exist.

Privacy

Privacy, of course, is not only what people make of it. The right of privacy has long been identified as a key component of human dignity and was described in *Griswold* v. *Connecticut* (1965) as a right "older than the Bill of Rights." Under the law, privacy in different forms has received protection under the First Amendment, which protects freedom of conscience; with the guarantees of the Fourth Amendment, which guard against unconstitutional searches and seizures; and with a host of statutory and common law protections. Despite these legal guarantees, the concept of privacy has also long been contested and debated as discontent over existing protections prompted debate and reconsideration of its significance in modern life, broadening its legal protections to include control over information, intimacies of personal identity, physical or sensory access, protection against searches and seizures, solitude, freedom from surveillance, and others. (Schoeman 1984).

This evolution of privacy is a result of many discussions of instances when the law as it stood did not entirely protect privacy as society thought it should. Then as now, technological innovations, changing conceptions of personhood, and shifting norms, values, and expectations have all challenged conceptions of privacy and directed changes in legal doctrine. The arrival of biometric technology may present novel challenges to existing notions of privacy, but it is important to recognize that this is an evolution, rather than a revolution, of an idea.

One of the most explicit, and most contested, guarantees of privacy is found in the jurisprudence of the Fourth Amendment. The drafters of the Constitution, concerned with searches by British soldiers, identified the first components of privacy: property and autonomy. As Slobogin (2007) explained, protection of possessions and persons translated into a privacy interest:

> When the government enters into one's house to seize possessions, property interests are most directly implicated. When the government seizes one's person, autonomy interests obviously come into play. But when the government engages in a search—i.e., when it "looks into" one's house or effects or "looks over" persons or their papers in "an effort to find or discover something"– the interest most clearly implicated is not property or autonomy but privacy. (24)

The Fourth Amendment's protection of a privacy interest was challenged by technological innovation and the governmental imperative to protect society against crime, both of which exacted costs on individual privacy. Although it is easy to see the rise of information technologies and the initiatives as a result of September 11 as unique risks to privacy, many cultural, political, and social changes have questioned the sociological and doctrinal boundaries of privacy.

One of the earliest legal discussions of emerging risks to privacy arose from the growing prevalence of the Kodak camera, which made it easy for anyone to take photographs. In 1890 Samuel D. Warren and Louis D. Brandeis described the diminution of privacy in the face of this technological innovation. They wrote, "Instantaneous photographs and newspaper enterprise have invaded the sacred precincts of private and domestic life; and numerous mechanical devices threaten to make good the prediction that 'what is whispered in the closet shall be proclaimed from the house-tops.'" The type of privacy that was threatened by modern enterprise was of a sort not yet recognized by the law:

> The intensity and complexity of life, attendant upon advancing civilization, have rendered necessary some retreat from the world, and man, under the refining influence of culture, has become more sensitive to publicity, so that solitude and privacy have become more essential to the individual; but modern enterprise and invention have, through invasions upon his privacy, subjected him to mental pain and distress, far greater than could be inflicted by mere bodily injury. (Brandeis and Warren 1890, 263)

In their seminal article, Brandeis and Warren declared that privacy was a fundamental aspect of natural law and an element of human dignity that provided for the "right to be left alone." Although fundamental, the right of privacy nevertheless had to be redefined in light of political, social, and economic changes that threatened the sanctity of privacy and left individuals unprotected under the law. Warren and Brandeis used the term *inviolate personality* to describe the right of the individual to decide which thoughts, sentiments, and emotions should be communicated to others. Their warnings did not go unheard. It was not long after the publication of their article that courts and legislatures began to carve out legal protection that attempted to fit their recommendations. The trouble, of course, was that the substance of privacy was a conundrum even though its value was accepted. Since this seminal article, many scholars, lawyers, and courts have attempted to lend substance to privacy, which has been continually modified in light of technology, science, and politics.

For example, nearly forty years after the publication of the article penned with Warren, in his famous dissent in *Olmstead* v. *United States* (1928), Justice Brandeis reasserted his defense of privacy, arguing that the protections of the Fourth Amendment should be greater. In this case, the Supreme Court was asked to consider whether wiretapping violated the guarantee against unreasonable search and seizure when the incriminating evidence was gathered by intercepted messages on telephones. The complicating nature of this "search and seizure" was that the insertion of the wiretaps was made without any physical trespass on any property of the defendants.

In his dissent, Brandeis echoed the language of his well-known law review article, describing privacy as "the right to be let alone—the most comprehensive of rights and the right most valued by civilized men." His dissent pushed the Supreme Court to account for the changing nature of privacy and its potential intrusions when he argued that "discovery and invention have made it possible for the government by means far more effective than stretching upon the rack to obtain disclosures in court of what is whispered in the closet." The logic of Brandeis's dissent was adopted in *Katz* v. *United States* (1967), when the Court held that the Fourth Amendment protected people, not places.

Yet Fourth Amendment jurisprudence was not the only place where the influences of Warren and Brandeis on discussions of privacy were felt. If privacy was fundamental to individuals and could not be set aside because of the machinations of society, politics, and technological innovations that cultivated new vulnerabilities and threatened "the immunity of person, the right to one's personality," then privacy was something over which the individual should exercise some control. For example, Warren and Brandeis wrote, "Common law secures to each individual the right of determining, ordinarily, to what extent his thoughts, sentiments, and emotions shall be communicated to others." This vision of privacy, conceived as an aspect of personhood, provided new philosophical grounds for debate and legal protection.

In *Griswold* v. *Connecticut* (1965), the Supreme Court lent legal substance to the idea of privacy as control over decision making about intimate matters. Justice William O. Douglas, considering the issue of state interference in intimate decision making, held that married couples have a constitutional right to privacy in the marital bedroom. He wrote, "Though the Constitution does not explicitly protect a general right to privacy, the various guarantees within the Bill of Rights create penumbras, or zones, that establish a right to privacy. Together, the First, Third, Fourth, and Ninth Amendments, create a new constitutional right, the right to privacy in marital relations." This decision marked a turning point in both the legal concept of the individual, who now possessed a "zone of privacy," and in subsequent cases limited the government's ability to interfere with decisions relating to marriage, contraception, and procreation.

Granting constitutional protection for privacy as a fundamental aspect of personhood created an ongoing debate about the substance of privacy and how to best protect it under the law. For example, William Prosser (1984) operationalized privacy, categorizing as offenses to privacy in intrusions on a person's seclusion, solitude, or private affairs; public disclosure of embarrassing facts; public disclosure of private persons in a false light; and appropriation. One of the preeminent privacy theorists, Alan Westin, suggested in his 1970 work that privacy is a necessary aspect of social life and has found a place in many different cultures in a variety of societal practices. Despite the centrality of privacy, vigilance was required to preserve it. Westin warned, "American society in the 1970's faces the task of keeping this tradition meaningful when technological change promises to give public and private authorities the physical power to do what a combination of physical and socio-legal restraints had denied to them as a part of our basic social system."

The changes about which Westin warned prompted other theorists to advocate legal protection for the individual to decide what affairs should be subject to public observation. Privacy now took on new philosophical dimensions and more legal complexity. Ruth Gavison (1984), for instance, held that privacy represented the irreducible elements of secrecy, anonymity, and solitude to protect the liberty of the individual "to establish a plurality of roles and presentation to the world" (365). As a function of individual choice of disclosure, the meaning of privacy expanded to a prohibition against observing individuals against their will, which afforded "a moral presumption to that preference" (Benn 1984, 19. Benn argued that autonomous individuals are empowered to make choices about privacy as long as there is not a compelling moral reason to supersede the choice. The imperative to weight claims of privacy against competing moral claims was familiar and reflected its Fourth Amendment origins when the obligation of the state to fight crime was pitted against the liberty interest of the individual. Yet the growing prevalence of information technologies enhanced the scrutiny of the individual far beyond the imagination of Warren and Brandeis. Surveillance of communications, surveillance of physical activities, and transactional surveillance increasingly were seen as an imposition on individual privacy, but these forms of information gathering also had their own set of moral imperatives and had to be balanced with individual privacy.

Whalen v. *Roe* (1977) illustrates the balancing act of moral interests when individual privacy is weighed against the societal interests in using personal information for the common good. In this case, patients and doctors challenged a New York statute that required copies of prescriptions for certain drugs to be recorded and stored in a centralized government computer, arguing that it violated their constitutional right to privacy. Although the Court rejected this claim, a majority of the justices argued that under certain circumstances, the disclosure of health care information may violate a constitutionally protected right to privacy. Justice John Paul Stevens identified informational privacy as one aspect of the constitutional right to privacy: "The cases sometimes characterized as protecting 'privacy' have in fact involved at least two different kinds of interests. One is the individual interest in avoiding disclosure of personal matters, and another is the interest in independence in making certain kinds of important decisions." Yet the interest in controlling information as a function of privacy was juxtaposed against the political objectives of protecting the common good and preventing harm. In *Whalen,* the Court was persuaded that the centralized database would foster an understanding of the state's drug control laws and ultimately further the "war on drugs." In recognizing the implications of the information being gathered, the Court considered not only the stated justification, but also whether the database represented a "useful adjunct to the proper identification of culpable professional and unscrupulous drug abusers."

The reasoning of *Whalen* continues to ring true. The idea that the individual has an interest in controlling information is not merely a subjective matter; it must be weighed against the information that society determines should be used for societal objectives. In considering the value of claims to privacy, it is necessary to remember that the state occupies a relative role in the individual's claim of privacy, serving to regulate against harms, making the idea of personhood a relational one—both positive and negative—rather than defined only as freedom from the state.

The competing interests that limit individual choice in the disclosure of information are not always those of the state and not always morally based; the interest might be commercial, for example. As Richard Posner (1984) argued, concealment or secrecy of information as a principle of privacy is often juxtaposed by competing interests in the information in both the public and private sectors, making individual privacy a consideration that must be weighed against business interests and compelling state interests.

Of course, the competition between individual privacy interests and the commercial and regulatory interests of the public and private sector has escalated because the gathering of information is now multiplied by a continually expanding set of objectives and facilitated by information technology over which the individual exercises little control. This puts individual privacy interests at a disadvantage. As James Rule (2007) aptly describes,

> Organizations are constantly finding new ways of capturing, transmitting, and using personal data, for purposes defined by the organization rather than by those depicted in the data. Supermarkets track our purchases; government agencies track our

travels, transactions, communications, and associations; insurers and employers monitor our medical histories and genetic makeup; retailers monitor our expenditures, our website visits, and our financial situations. (13)

Information technology has increased the presence of physical, communications, and transactional surveillance and takes place with little prohibition against the sharing of information between the public and private sectors. The privacy debate is again front and center. The question is whether these forms of information gathering—to protect against credit card fraud, enhance law enforcement capabilities, dole out benefits, or create advertising—equate with the demise of privacy and the alteration of our modern day-to-day life to worse ends. "Our digital biographies are being assembled by companies which are amassing personal information in public records along with other data. Collectively, millions of biographies can be searched, sorted, and analyzed in a matter of seconds" (Solove 2004, 47).

What are the implications? Some argue that technological and social changes in the public and private sectors are stripping the value of privacy from individuals. Barry Steinhardt, director of the Technology and Liberty Project at the American Civil Liberties Union, has argued that technology constitutes a "monster" that needs to be restrained:

The explosion of computers, cameras, sensors, wireless communication, GPS, biometrics, and other technologies in just the last 10 years is feeding what can be described as a surveillance monster that is growing silently in our midst. Scarcely a month goes by in which we don't read about some new high-tech method for invading privacy, from face recognition to implantable microchips, data-mining to DNA chips, and now RFID identity tags. The fact is, there are no longer any technical barriers to the creation of the surveillance society. While the technological bars are falling away, we should be strengthening the laws and institutions that protect against abuse. Unfortunately, in all too many cases, even as this surveillance monster grows in power, we are weakening the legal chains that keep it from trampling our privacy. We should be responding to intrusive new technologies by building stronger restraints to protect our privacy; instead, all too often we are doing the opposite.[1]

Consider the warning of philosopher Jeffrey Reiman (2005) in commentary on the prospects of information gathering: "When you know you are being observed, you naturally identify with the outside observer's viewpoint, and add that alongside your own viewpoint on your action. This double vision makes your act different, whether the act is making love or taking a drive" (xx). Consider as well that Americans believe their right to privacy is under serious threat. According to one survey, 64 percent of people have "decided not to purchase something from a company because they weren't sure how their personal information would be used" (CBS, 2005).

Despite the cataclysmic predictions some have made, disagreement is growing about whether privacy is in fact in peril. There are indications that people are willing to provide personal information rather readily. For instance, some have shown that individuals are actually less concerned about privacy than might be expected and will trade personal information for seemingly small rewards, including small sums of money (Spikerman, Grossklags, and Berendt 2000). In line with this trend, Acquisti (2004) has demonstrated that individuals engage in online interactions despite the potential risks:

> First, given that the individual loses control of her personal information and that information multiplies, propagates, and persists for an unpredictable span of time, the individual is in a position of information asymmetry with respect to the party she is completing a transaction with. Hence, the negative utility coming from future potential misuses of off-line personal information is a random shock practically impossible to calculate. Because of identity theft, for example, an individual might be denied a small loan, a lucrative job, or a crucial mortgage.
>
> In addition, even if the expected negative utility could be estimated, I put forward the following hypothesis: when it comes to security of personal information, individuals tend to look for immediate gratification, discounting hyperbolically the future risks (for example of being subject to identity theft), and choosing to ignore the danger. Hence, they act myopically when it comes to their off-line identity even when they might be acting strategically for what relates to their on-line identity.[2]

Perhaps the answer lies somewhere in between. Clearly, personal information is increasingly a currency that supports our interactions in transactions, communications, and social networking, and it does not follow that privacy is in peril. Today the issue is not only about limiting intrusions; it is also about enabling interactions. Our society makes use of technology to facilitate communications, complete transactions, or for entertainment purposes. Individuals increasingly make choices to share thoughts, sentiments, and emotions using technological engagement. While privacy advocates have long advanced a strategy of data minimization so that government and corporations are less able to abuse personal information, Peter Swire (2009) has described an emergent ideology of "data empowerment," where people control information about themselves through online social networking and other sites. For example, in 2007the Pew Internet & American Life Project explored questions of teen online privacy "by looking at the choices that teens make to share or not to share information online, by examining what they share, by probing for the context in which they share it and by asking teens for their own assessment of their vulnerability."[3] The results demonstrated that many youth actively manage their personal information, balancing confidentiality of important pieces of information with the process of creating content for their profiles and making new friends at their discretion.

This does not mean, however, that privacy ceases to be an essential element of our politics, morality, society, and culture. The question, especially with regard to emerging technologies such as biometric identification, is to understand when and how privacy matters to individuals so that technological development and policy initiatives reflect and protect values that are important to individuals, including privacy. Understanding the societal values of privacy and the perception of vulnerabilities to it are the starting points for coming to terms with how much and in what manner privacy has been challenged and changed. Societal perceptions are important forces of transformation, causing technology, science, and politics to evolve in ways that are reflective of long-standing ethical, political, and legal imperatives but also indicative of emerging trends.

Normative Perceptions of Privacy

To the end of understanding the value of privacy and its vulnerabilities, the qualitative focus groups for the research were designed to discern the views of biometric users and nonusers on a variety of issues related to how personal information is protected and threatened in different institutional contexts and to develop an understanding of how biometric technology figures into the mix.[4] Not surprisingly, the issue of controlling information is key to understanding the effect of biometric technology and the information it generates in relationship to privacy. In focus group discussions, individuals consistently expressed a concern of losing control over personal information, though it did not necessarily limit their use of personal information for a range of transactions. For example, the focus group participants discussed privacy as a form of secrecy or confidentiality that they controlled, sharing it with other individuals or trusted confidents. As one person noted, "Privacy is the ability to keep information to yourself, the relevant parties, and sometimes to potentially relevant parties, by controlling the divulgence of one's information."

Privacy in this context is conceived of as control over personal data or control of what others (individuals, companies, or government) know about the individual. Collection of information by businesses was viewed as an invasion of privacy by some participants, but it was also accepted as part of what was necessary to participate in transactions, networking, or communication. Some forms of sharing information were considered more dangerous than others. For example, the collection of information by retailers who offer sale cards at the point of purchase was considered a serious invasion of privacy. One nonuser even refused to use these cards because of the privacy threat. Also, one user suggested that since debit card use has become widespread, sales clerks rarely check to see if a credit card signature matches the name and signature on the card. As one user commented, "Point of sale is always a weakness."

One user suggested that the availability of private information is not by itself dangerous. Rather, how the information might be used creates the danger: "It's what you know how to do with that information. . . . If I came across a page of social security numbers, I wouldn't know what to do with it."

Participants also offered some additional comments on how safe their information is, mentioning both hacking and identity theft as threats to their security. Most believed that their personal information is vulnerable, listing examples of cases in which an employee at an institution responsible for collecting personal information made improper use of it. However, fears about hacking were tempered for some by the opinion that hackers tend to choose big targets and do not bother with individuals who have modest assets.

Most participants, however, regarded identity theft as a serious threat. They noted that they provide personal information even to institutions whose trustworthiness they are unsure of because they have no choice. They cited applying for a job as a situation where they are often required to provide personal information. Participants commented:

"I think anybody who wants your information can get it. . . . If you try hard enough you can get anybody's information."

"They [insurance companies] probably share a lot of information and put all the records together."

"For me, I can't be 100% sure that my information won't be used without my authority."

In discussing the extension of "privacy" as the sharing of personal information, focus group participants agreed that individual concerns were moderated by the social elements of enhanced security, especially in the wake of terrorist activity and changes attributable to the increase in the exchange of personal information as part of the information age. Individuals also admitted that they acquiesced to the use of personal information in their transactional relationships with public and private entities because with and government increasingly require the use of personal information even in light of threats to privacy.

The national survey bore out similar trends. In this survey, respondents were asked to answer a series of questions regarding their understanding of privacy as it related to personal information. The majority of participants reported being very protective of their information and were at least minimally knowledgeable about how the government handles the privacy of such information. As was the case in the focus groups, the results of the national survey also indicated that protection of personal information was an important concern related to privacy. The respondents reported that the protection of their personal information was very important to them: 89.1 percent reported it was extremely important, 9.1 percent reported it was somewhat important, while only 1.4 percent reported that it was not important to them.

The protection of personal information as a value of privacy requires an understanding of the vulnerabilities that threaten it. The highly publicized breaches in data security in the cases of Choicepoint and Lexis Nexis, reported increases in identity theft, and the coupling of the war on terror with an increased use of personal information and surveillance have led to a sense that while personal information is necessary as currency of interactions in the public and private sector, there are vulnerabilities in data security. To explore the perceptions of risks relative to the increasing use of personal information in the public and private sectors, respondents in the national survey were asked to indicate their level of concern with a number of different threats to data security that could affect them personally (e.g.,

Table 3.1

Average reported levels of concern with various threats to data security (N = 1,000)

Issues	Mean (sd)
Unwanted telemarketing	4.2 (1.2)
Unwanted e-mails or spam	4.3 (1.2)
Identity theft	4.5 (0.9)
Someone revealing your personal information to government without your permission	4.3 (1.1)
Someone revealing your personal information to private companies without your permission	4.6 (0.8)
Someone getting into your bank accounts without your permission	4.6 (0.9)
Loss of our civil liberties as part of the war on terror	4.1 (1.3)

Note: There were some statistically significant differences in regard to respondents' gender, political affiliation, education level, and age. See appendix B.

someone revealing personal information to the government about them without their permission). They indicated their level of concern on a scale of 1 to 5, with 1 representing "not at all concerned" and 5 representing "very concerned." Their average levels of concern are presented in the table 3.1.

The prevalence of telemarketing and spam pointed to concerns related to intrusions on privacy as solitude, but also indicated a risk of the potential misuse of personal information for marketing or promotion. Similarly, identity theft posed the risk of a loss of control over personal information, damage to reputation, and the disruption and devaluation of the currency of information. Revelation of personal information to the government by private corporations also provoked concern, as did the loss of civil liberties in the war on terror. Vulnerabilities of data to security breaches in the public and private sectors are juxtaposed against an increasing reliance on information technology. Perceptions of risks and vulnerabilities point to perceived losses of privacy relating to personal information, and anticipating, identifying, and combating them can help to inform technological and policy solutions. As Flanagan, Howe, and Nissenbaum (2008) noted:

> If an ideal world is one in which technologies promote not only instrumental values such as functional efficiency, safety, reliability, and ease of use, but also the substantive social, moral, and political values to which societies and their peoples subscribe, then those who design systems have a responsibility to take these latter values as well as the former into consideration as they work. (322)

All of these perceived vulnerabilities were significant threats to the value of privacy, but does it follow that minimizing data that we distribute is the answer? Given the reality of relying on personal information as currency for our interactions, it seems impossible to put the genie back in the bottle, making the objective of protecting the value of privacy a different one: to empower individuals to use personal information as currency while minimizing the risk. The answer lies in protecting the use of personal information. But the protection of personal information as currency involves a paradigm that does not rely on privacy for protecting individual liberty. Because our privacy interest in personal information is used as a currency for our interactions, protection of it requires a consideration of trust and confidence in public and private sector institutions that handle personal information, an assessment of policy objectives to which the personal information will be put, and a continued concern for the decisional autonomy of the individual in the development and deployment of technology. The liberty interest that might once have been termed only "privacy" is now a multifaceted approach that enhances the value of personal information as currency rather than the futility of arguing only for data minimization. The first step of protecting personal information, however, is to conceptualize it in a way that facilitates policy and technological safeguards.

Personal Information as Property

Many have framed the debate about privacy and personal information as dominated by analogies to property and intellectual property rights (Cohen 2000). The idea of property as applied to personal information attempts to apply the ideas of ownership, liberty, and choice to personal data. The notion of property, however, is not an exact semantic fit. The notion of privacy has long been tied to property and the individual's ability to exercise control by its possession, thereby creating the conditions of liberty. "If owners have a power to exclude in a private property system and economy, then they can establish and protect various personal goods. Among them are such overlapping items as autonomy, personality, self-respect, self-esteem, liberty, control privacy and individuality" (Munzer, 1990, 90). The philosophical starting point for the connection between liberty and property is John Locke, whose views on property were defined by the importance of mixing one's labor with unowned things and the expected benefit that follows. In this sense, property provides the basis for self-determination.

Conceiving of personal information in this manner is imperfect for several reasons. The property "interest" in personal information is complicated by the confusion created by the usual Lockean fruits-of-labor justification of property ownership. Personal information, which pertains to the individual, is created not by the individual but by competing interests, whether they be governmental or commercial, to serve a purpose. Although the personal information may come within the property interest of the individual, it also is controlled by the governmental or commercial entities that created it. The social security number for the purpose of social security benefits; the drivers' license for the purpose of driving a car; the credit card for the purpose of a convenient transaction: all of these circumstances represent instances where the personal information refers to an individual but has not been created by an individual. Also, the analogy of ownership of real property to ownership of personal information is difficult to overcome. As Pamela Samuelson (2000) points out, prior conceptions of property have not generally included personal information: "Although the law often protects the interests of individuals against wrongful uses or disclosures of personal data, the rationale for these legal protections has not historically been grounded on a perception that people have property rights in personal data as such. Indeed, the traditional view in American law has been that information as such cannot be owned by any person" (1125).

Individual choice is also difficult to ensure in the paradigm of personal information as property. According to this market model of regulation, an individual's interest in personal information is traded by choice, with either notification or enrollment. Here individuals decide when and how to trade their personal information, and by implication their privacy, because the regulation of the exchange would raise exchange costs and undermine the market. The idea of freedom and choice in relationship to property in terms in a democracy and at first glance seems a viable means to negotiate exchanges of personal information. Generally, however, it is the vendor or the government that establishes the terms of the exchange, and even if the consumer or constituent is aware of the proposed uses, there is not a realistic element of choice. Individuals cannot fully understand or contemplate the uses to which their personal information will be put, and thus the notion of choice is rendered inadequate. The idea is that eventually the market will account for the preference of individuals to protect their information, and commercial entities subsequently will build in greater protections, both technological and policy, to match consumer preferences.

The problems with this approach are obvious. There are many circumstances, especially with regard to projected uses of biometric identifiers, when the element of choice is at best truncated. Do not submit a fingerprint: do not drive. Do not acquiesce in enrolling in an employer's identification system: do not work. Similarly, although it is possible to consider the divulgence of personal information to a credit card company to be a matter of choice, participation in the marketplace without the benefit of some exchange of personal information may be an increasingly difficult proposition. Although the divulgence of personal information as a choice model is designed to empower consumers in the marketplace of privacy, much would have to be done to accomplish this end.

Another complicating factor of personal information as property is the fact that alienability of personal information is difficult to limit. Unlike usual transfers of property, the transfer of personal information is never quite complete. That is, the individual does not completely divorce himself or herself from personal information. It can, in other words, be traded again by anyone who acquires it. As Pamela Samuelson (2000) and others have pointed out:

> It is a common, even if not ubiquitous, characteristic of property rights systems that when the owner of a property right sells her interest to another person, that buyer freely can transfer to third parties whatever interest the buyer acquired from her initial seller. Free alienability works very well in the market for automobiles and land, but it is far from clear that it will work well for information privacy. (1125)

The ownership of property interests in personal information goes hand in hand with a utilitarian market approach that has been criticized for its information asymmetry resulting in market relationships that are skewed. Cohen (2000) explains:

> The data privacy debate is also a debate about freedom of choice and its necessary preconditions. The prevailing approach to this question is closely aligned with the position that personally-identified data becomes "property" when, and because, it becomes tradable: A successful data privacy regime is precisely one that guarantees

individuals the right to trade their personal information for perceived benefits, and that places the lowest transaction cost barriers in the way of consensual trades. If individuals choose to trade their personal data away without placing restrictions on secondary or tertiary uses, surely it is their business. On this view, choice rather than ownership is (or should be) the engine of privacy policy. What matters most is that personal data is owned at the end of the day in the manner the parties have agreed. (1391)

The utilitarian approach to managing privacy assumes that society is better off if the preferences of most of its members are satisfied. The problem with applying this theory to the management of privacy is that the incentive structure for handling information is slanted to the interests of the public and private sector institutions acquiring the personal information. The individual is not always apprised of the full range of uses to which the information might be put, so the interest of the individual in retaining control over his or her personal information stands uneasily next to the collection and collation of information by government and commercial entities. Although there are surely instances when the interest of the individual in divulging his or her information coincides perfectly with the collection of the information for some good, transaction, or benefit, there also may be instances when the individual has little understanding of how much personal information has been taken, to whom it has been given, or what benefit he or she, or others, have gained. In this instance, the property interest of the individual begins to work at-cross purposes with that of the vendor of information (Mercuro and Medema 1997).

The difficulty of conceiving of personal information as property raises a host of problems, not the least of which is that impinging on property ownership is viewed as an infringement on a certain type of liberty. The conveyance of liberty by a property right is accomplished by allowing the individual a modicum of control over his or her tangible interests and the process of self-actualization that follows from it. If a right to privacy is defined by a property interest in personal information, it may be that it is not only the idea of property that is limiting but also the idea of a right that is limiting. The language of rights pertains to the ideal that they seek to promote. At a basic level, the idea of property rights is directed toward the protection and promotion of ownership and the maximization of benefit with a system of utilitarianism.[5] Despite the conceptual limitations associated with property rights, it does not follow that we cannot explore the underlying social goals of such a system for protecting personal information. For instance, the notion of property rights was conceived to balance "the distribution of power and the prevailing patterns of motivation in the societies in which we live" (Scanlon 2003 , 36. The idea of liberty and efficiency, both implicit in the protection given to property as an individual right protected in a utilitarian framework, can certainly be conceptually extricated to reconceptualize the right of privacy as it pertains to personal information. If the basic philosophical foundation changes, so too must the ideas of costs and freedoms promoted with the advocacy of rights (Cohen, 2001).

For example, the benefit of speaking about privacy interests as an individual right, separate from a property right, is obvious. A legitimacy is conferred with the notion of rights, and surely it is difficult to set aside the legal, political, and moral weight of rights. As an instrument, rights can serve as "trumps over some background justification for political decisions that states a goal for the community as a whole" (Dworkin 1984, 153). While some argue that rights-based moralities are impoverished, the assertion of a "right" to individual privacy is not one that is limited to its positive law definition.[6] As has been pointed out, there is a disconnect in liberal theory that portrays individual rights as grounded in positive law and that debates about natural, moral, or human rights go beyond, or exist as background to the rights of a legal system (Waldron 1984). Yet is this distinction reflected in the normative understanding of the right to privacy? The assumption that the interpretation of positive law results in an artificial separation between the law and the normative concerns is misguided. Instead, the idea of the "right of privacy" reflects institutional arrangements and moral justifications within which the right it situated. The right can still be a trump, but the question is, A trump against what? The notion of personal information is not merely a discussion of rights or of property but instead a debate about the conditions under which the protection of personal information as a social goal exists, taking normative judgments of duty and obligation as indications for the development of policy.[7]

Privacy as Social Freedom

An alternative to a privacy regime based on the right of property is one that conceptualizes a right to privacy not as isolation but as a dimension of social freedom. This idea of privacy fits well with the direction that the information revolution is taking as privacy is increasingly not limited to protection against intrusion but also translates into empowerment for individuals. This definition allows for the reconciliation of the usual tension that arises between privacy as an individual right and a competing societal interest. Regan (1995) noted that a competing interest with respect to privacy might be a "police interest in law enforcement, the government interest in detecting fraud, and an employer's interest in securing an honest work force" (225). If privacy is always seen as an opposite to societal objectives, whether governmental or as part of the fabric of social interaction, then the accommodation of each is a difficult task, especially as it relates to personal information. It may be helpful then to think about the "right to privacy" as not limited by its spatial definition and its conceptual connection to visibility.[8] This conception of privacy does not stand in opposition to the attainment of a benefit or a common good, whether it is commercial or governmental, but instead must be a factor in contemplation of policy. The idea of privacy cannot be completely understood through the lens of an atomistic individual. Privacy, in this slightly ironic sense, is, and contributes to, the social organizations that are the constituent components of a common good upon which the individual depends. As Regan (1995) suggested, "This lack of acknowledgement of social organizations and the role they play in the latter twentieth century relegates privacy to not only a narrow sphere but also one that may be obsolete" (218). This criticism is leveled at a notion of democracy that hinges on the idea of the isolated individual without recognizing the important role the public sphere plays in the private life of the individual , with each dependent on the other. This insight was one recognized by John Dewey (1954): "Yesterday and ever since history began, men were related to one another as individuals. . . . Today the everyday relationships of men are largely with great impersonal concerns, with organizations, not with other individuals. Now this is nothing short of a new social age, a new age of human relationships, a new stage setting for the drama of life" (96).

Changing the idea of privacy from a right to isolation to protected participation in social life—networking, communications, transactions—accounts for individuals engaging in interactions, by choice and by necessity, that require the divulgence of personal information without viewing it as a requisite loss of privacy. If we understand the terms of individual engagement that serve to create the conditions of social freedom through the exchange of personal information, then we can better protect it. Individual interest in privacy of personal information does not entail girding ourselves from the authority of the state or commercial entities; instead, it involves the ability to frame our interest in personal information, not only necessarily as a violation of privacy but as a means of participating in various associations and allowing a variety of competing values, such as security, protection against identity theft, or ensuring identity assurance in transactions.

The most important part of this change in thinking about privacy is that the use of personal information is not in all cases an invasion to be staved off. In fact, the social world now depends on individuals who participate, and this participation increasingly includes the provision of personal information. This may explain the seeming contradiction that Americans view their privacy as under threat but nonetheless continue to interact by freely exchanging the currency of personal information; it may not be a contradiction but a fact of a social life increasingly built with information sharing. The trend toward data empowerment reflects the idea that participation in the social world can be a validation of selfhood: "The transition from privacy as an experienced phenomenon to privacy as a normative validation of selfhood is made by postulating that persons perceive the social practice of privacy and take it to be confirmation of their status as a moral agent" (Weinrub 2000, 40).

Yet to ensure participation in the social world as an autonomous agent, the moral agency of the individual must be protected and respected in policy and technological design. Respect and protection of choices involving the disclosure and use of personal information give value to the idea of autonomy, which is part and parcel of the normative idea of privacy. This respect for autonomous moral agency is also important for limiting the imposition of the public sphere in an age of information. "In George Orwell's 1984, Winston is brought to Room 101 because everything, even his unexpressed thoughts and emotions, is deemed to be a proper object of concern to, and at the service of, the state" (Weinrub 2000, 32). Without an eye toward protecting the moral autonomy of the individual, social freedom quickly becomes social oppression.

To put it simply, the question is not whether personal information should be shared but how it should be protected. The use of the currency of personal information in communications, interactions, and transactions carries a risk, but it is no longer realistic to require that individuals refrain from using personal information. A reconceptualization of privacy of personal information from an isolated right to one that is the currency of social freedom requires that the rubric of privacy be expanded to include a consideration of the values, technologies, policies, and institutions that are implicated in the sharing of personal information. The language of a "right of privacy" as it pertains to personal information as a form of data empowerment must be modified to include new terminology that better describes the preservation of individual liberty and decisional autonomy, the elements of trust and confidence of institutions that increase the accountability of personal information, and justifications for and limitations on technologies that regulate human behavior.

Biometric technology is at the center of these debates. Biometric systems of identification can provide identity assurance and offer a potential form of protection in the expanding universe of personal information. But they also rely on personal information, triggering concerns of abuse and misuse. It is necessary to understand an individual's interaction with biometric identification systems to gain a better sense of the complicated nature of values that are implicated. Defined broadly, the points of collection, processing, and dissemination present potential pitfalls for societal acceptance, requiring societal assessment of biometric systems of identification to go beyond the type of information that is collected. If biometric identification systems are to gain societal acceptance, the calculation must include the preservation of the norms, values, and expectations of society. This includes building trust and confidence in the institutions with which individuals are exchanging their personal information and recognizing that the values of individual liberty are relational and limited by concerns of the common good and its objectives. In short, considerations of institutional accountability, the legitimacy of policy objectives, and the protections for individual liberty figure into societal acceptance of biometric identification systems. Of course, none of these considerations is entirely new, but they have changed, and the societal acceptance of biometric technology rests on them. They deserve renewed debate.

4 Anonymity

The space in which we live, which draws us out of ourselves, in which the erosion of our lives, our time and our history occurs, the space that claws and gnaws at us, is also, in itself, a heterogeneous space. In other words, we do not live in a kind of void, inside of which we could place individuals and things. We do not live inside a void that could be colored with diverse shades of light, we live inside a set of relations that delineates sites which are irreducible to one another and absolutely not superimposable on one another.

—Michel Foucault, *Des Espace Autres* (1967)

Anonymity and Decisional Autonomy

Privacy as it relates to personal information and the presence of surveillance and information technologies must be modified by the principles of anonymity and decisional autonomy. This is because privacy is limited by the markers of space and visibility.[1] In contrast, the idea of anonymity has served an important purpose in preserving some degree of private space for political opposition and the expression of ideas, while decisional autonomy has been a necessary condition for fostering participation in political life, allowing the individual to make choices about associational and private life. Julie Cohen (2000) identified the cornerstone of a democratic society as

informed and deliberate self-governance. The formation and reformation of political preferences—essential both for reasoned public debate and informed exercise of the franchise—follows the pattern already discussed: Examination chills experimentation with the unorthodox, the unpopular, and the merely unfinished. A robust and varied debate on matters of public concern requires the opportunity to experiment with self-definition in private, and (if one desires) to keep distinct social, commercial, and political associations separate from one another. (1427)

Some view technologies that accumulate personal information about our physical activities, transactions, and communications as a potential threat to anonymity and decisional autonomy, chilling the expression of ideas and constraining associational activity with observation. Biometric systems of identification often fall under this criticism because they can potentially be used to track human behavior. Yet although regulation of human behavior can be a consequence of surveillance technologies, control is beneficial in certain circumstances. Long before the current debate about technological surveillance, Jeremy Bentham recognized the potential of architecture.

Bentham realized the effect of the power of observation on individual behavior in his description of a panoptic tower in the center of the reformatory where prisoners were under the influence of observation, even if they were not actually being observed. Bentham pointed to the self-disciplinary effects of observation on individual behavior as a technique of the modern state. This concept was appropriated by Michael Foucault (1995), who described the normalizing and constraining aspects of the power of observation, which served to limit the subjectivity, and thus liberty, of individuals: "He who is subjected to a field of visibility, and who knows it, assumes responsibility for the constraints of power; he makes them play spontaneously upon himself; he inscribes in himself the power relation in which he simultaneously plays both roles; he becomes the principle of his own subjection" (202–203).

Foucault extended Bentham's insight to the effects of the formation and accumulation of knowledge in modern bureaucratic institutions. These springs of bureaucratic power imposed regimentation and self-discipline on the subjectivity of modern individuals. The panoptical observation with technological and omnipresent observation evokes a vision of ultimate social control. The crux of the argument regarding the effect of surveillance on society and culture rests on the disembodiment and dissociation that occur with technological observation. The undetectable and unobtrusive quality of technological observation is not readily perceived as in the physical world of social and political interactions, making its effects less apparent but perhaps more detrimental in regulating human behavior. The constancy of observation results in self-imposed limitations on mobility and action due to the presence or even the threat of surveillance, causing individuals, prompted by the fear of being catalogued or tracked in social and political space, to change their behavior (Agre 2001). This criticism has been extended to the broad spectrum of surveillance technologies in the information age, all designed to facilitate the capture of personal information in the form of demographics, communications, transactions, or simply physical movement. September 11 served to refocus the concerns regarding the social and cultural consequence of surveillance technologies.

Foucault's contribution to the study of surveillance is significant; however, his simplistic appropriation is not sufficient to model the complexity of institutions and rationales that provide the context for modern surveillance. As Lianos (2003) discussed,

> The first step towards a less biased approach of Foucault's contribution on control would be to focus on a consistently disregarded point: that the Foucauldian model of control, and consequently its explanatory power, refers to the past and is not concerned with the emergence of the contemporary postindustrial subject. This is almost self-evident but often put aside in order to make things easier for today's analysts and to project Foucault's thesis on modernity unaltered onto the late capitalist present. Such unconscious blindness also disregards the fact that the analysis of control was for Foucault yet another axis of a clear cross-cutting theme in his research: the constitution of the human being as subject. However, things have advanced too far to reverse all that is based on this arbitrary reasoning. Only a model which takes the risk of referring directly, and not by analogy, to the contemporary condition, can supply the debate on control with a sound basis for addressing the questions of the late modern environment. (413)

Applying Foucault's insights to contemporary surveillance requires accounting for the complexity of institutional uses and the social processes associated with the dynamics of observation of physical activities, transactions, and communications. The growing prevalence of surveillance technologies in both the public and private sectors has served to fragment the dynamics of power that Foucault addressed, necessitating a more contextual understanding of institutions and their social processes.

Given this fragmentation, it is also important to recognize that the constitution of power is not unilateral. The individual is not merely an object under constant observation by one entity. To argue this is to discount the human agency. Instead, the individual is an actor who interprets and reacts to surveillance and, in some cases, benefits from it when the object is preventing fraud or detecting terrorist activity. The control effected by observation is not undifferentiated, yielding power only to those who are the observers. The power of observation is dependent on the perception of and response to the observation rather than an absolute and universal calculation of its effect. The individual may also exercise some amount of control by virtue of being aware, conscious, accepting, or manipulating of surveillance technology. For the purpose of understanding the effect of surveillance technologies on society, the question of conscious perception is significant. Majid Yar (2003) argued, "What this implies is that the functioning of panoptic power rests in its essence not upon visibility (the fact that the subject is visible to the eye that observes), but upon the visibility of visibility i.e. that conscious registration of being observed on the part of the subject (seeing and recognizing that he is being seen) is what induces in the subject the disciplining of his own conduct" (261).

The differentiation of the power of observation informs the power dynamics between surveillance technologies and those under surveillance. Those subject to observation may be conscious or unconscious of the surveillance and accepting of or resistant to its use. The role of human agency must be accounted for in the complexity of human and nonhuman (technological) actors. The language of description by those deploying the technology is as consequential as the depiction of those subject to it. As Latour (2007) argued, "The fact that no structure acts unconsciously 'under' each speech act does not mean that it is made out of thin air by 'local' linguists stuck in their office. It means that the written structure is related, connected, associated to all speech acts in some ways the enquiry should discover" (177).

John Gilliom's analysis of single welfare mothers who are required to participate in the Client Registration Information System–Enhanced (CRIS-E) presents an example. Gilliom (2001) identified CRIS-E as "a computer system that manages welfare cases, implements regular number matches and verification programs, and issues automatic warnings and compliance demands for all of the state's welfare clients" (1). According to Gilliom, the power of surveillance over the lives of the single women he studies is overwhelming. The welfare recipients described the effects of this surveillance as a "prison" and that a "slip-up" could ultimately cause them to lose their welfare benefits. The welfare mothers' perceptions of the surveillance technology reveal much about the sociopolitical context that influences their perceptions of the technology and of themselves. Gilliom quotes one mother as describing her interaction with welfare surveillance as "someone watching like a guard. Someone is watching over you and you are hoping everyday that you won't go up the creek" (1).

The perception of the surveillance by welfare mothers is indicative of how they see themselves in relationship to the welfare agency, but these perceptions can also be reflective of a larger social and political context. The welfare mother's reaction to surveillance depicts a framework of needs, care for her children, and the demands of her life and how surveillance technologies impede them (Gilliom, 2001). These descriptions of surveillance technology reflect a loss of personal control and autonomy in relationship to the bureaucracy of the state as it attempts to control fraud and manipulation of the welfare system. These descriptions also reflect the rhetoric used to describe the technology (and those under its purview), which draws on implicit ideals of the relationship between state and individual to both critique and justify the use of surveillance. Yet there is much more to the description of the surveillance technology than first appears. As Margaret Little (2003) attempted to demonstrate in her study involving women on welfare in Ontario, the politics of justifying the treatment and surveillance of poor, single mothers plays on certain myths or characterizations that are convincing and thus legitimize observation. In her study, she found that the depiction of welfare mothers as lazy helped to popularize support for their treatment. The characterization of welfare mothers as a drain on society ultimately swayed popular opinion toward the acceptance of surveillance of this group.

The rhetoric of justification of the use of surveillance technologies draws on and reflects the political ideals that are salient, influencing the perception of the surveillance. The rhetoric of justification also affects the interpretation given to the use of technology, providing a social construct for individuals to ascribe meaning to surveillance and, because of the fragmented nature of it, accept its use in certain institutions, on certain targeted groups, and in various societal contexts. This is particularly true with biometric systems of identification. The perceptions of the surveilled and the rhetorical justifications behind the deployment of biometric technology are important to understanding the beliefs and values that contribute to the social acceptance of biometric technology. Biometric systems of identification, because of their diverse institutional uses and the strong rhetoric of rationalization, serve as a useful device to capture the dynamics of acceptance and rejection of surveillance within the broader ideological and legal framework of anonymity and decisional autonomy, both of which serve to create freedom of association in civil society. Without the preservation of anonymity and decisional autonomy as elements of liberty, freedom of association in civil society suffers, but it does not follow that there is not a place for surveillance technologies in modern civil society.

The notion of civil society is not completely devoid of state power, and in fact, the presence of state power is a necessary component of preserving civil society. "Citizenship is one of the many roles that members play, but the state itself is unlike all other associations. It both frames civil society and occupies space within it. It fixes the boundary conditions and the basic rules of associational activity (including political activity)" (Walzer 2003, 318). As Walzer points out, after the dismantling of totalitarian states in Central and Eastern Europe, freedom did not exist; chaos did. A civil society requires an acceptance of state power to redistribute and manage resources and to protect property and civil rights. Some state control, in the form of surveillance, laws, or police power, must exist in order to create the conditions of security that allow the freedom necessary for associational life to flourish. An understanding of how surveillance technologies affect anonymity and decisional autonomy in a civil society increasingly built on the exchange of personal information must consider the implications of a world, and state power, altered by the information age and September 11.

Participation in the social world requires the integration of personal information in our interactions within civil society but also with state authority that makes use of technological tools to combat crime, fraud, or terrorism. It may seem contradictory to assume that the increasing attendance of surveillance technologies in our lives does not translate neatly into a loss of anonymity and decisional autonomy. Yet the fragmented institutional use of surveillance technologies and the interaction between the actors and nonactors in an ever increasingly wired world has taken the "Big" out of anthropomorphic "Big Brother." The introduction of these complexities is not meant to minimize the social, cultural, and political implications of surveillance technologies. Rather, the accommodation of information technologies in the more traditional notions of state and civil society involves a reconsideration of the nature of state authority and the medium of civil society, components dependent on and constructed with technological and informational means.

There are also considerations of degree. The specter of constant surveillance can certainly create the sense in individuals that they are not able to freely associate, thereby chilling the vibrancy of civil society. Walzer (2003) writes that "the words 'civil society' name the space of uncoerced human association and also the set of relational networks—formed for the sake of family, faith, interest, and ideology—that fill this space" (306). The relationship between anonymity and decisional autonomy and surveillance in civil society is more a question of balance than it is choosing one over the other. The challenge is to account for the usefulness of surveillance technologies without allowing the presence of them to devolve into a threat against the very thing it is meant to protect. Despite the novelty of some surveillance technologies, the dilemma of balancing state authority and civil society is not new. The potential effects of surveillance technologies must be understood in the context of a long legal and philosophical tradition against which the contemporary uses of surveillance can be juxtaposed. Historical context is helpful in defining points of continuity and change.

Historical Antecedents: Anonymity

Anonymity has long been associated with First Amendment freedoms, but its theoretical foundations are preconstitutional. In this sense, the purpose served by anonymity was not indicated by an absence of identity but as a mask for identity (Griffin 2003). The mask of identity served a variety of purposes and was tied in its early development to the written word. The practice of associating authors' names with text predates the Middle Ages (e.g., Caesar's *The Gallic Wars*), but anonymous writing remained common even with the evolution of efficient printing and distribution methods. In the English legal tradition, printers were generally held responsible for the works they published, while authorial anonymity was tolerated. Writers, depending on the purpose of their authorship, might choose to be anonymous for many reasons: ethical, religious, social, commercial, for self-effacement, or to avoid the distraction of persona. Anonymity relieves anxiety over exposure, fear of persecution, or political challenges, and it can satisfy hope for unprejudiced reception or the desire to deceive by inventing authority where it may be lacking (North 2003).

The political history of anonymity is often associated with the use of pseudonyms in the debates between the Federalists and the Anti-Federalists. For the Federalists and Anti-Federalists, anonymity in political writings was important protection for unpopular viewpoints. The Federalist Papers and the Anti-Federalist counterparts debated the future composition of the constitutional framework while remaining under the cover of pseudonyms, including Publius, Cato, Centinel, The Federal Farmer, and Brutus. In fact, during the transition from the Articles of Confederation to the framing of the Constitution, newspapers were replete with discussions from the Anti-Federalists and the Federalists. The founding fathers fiercely defended anonymity as a necessary foundation for a free and democratic society. In the context of the debates, anonymity protected against recrimination because of political association or belief. The use of pseudonyms to protect anonymity also found a place in the debates of the abolition movement as those who criticized the institution of slavery protected themselves against mounting political tensions.

The interest in protecting against recrimination for political beliefs and association informed the decisions of the U.S. Supreme Court that protected political speech and association. For example, the protection against disclosure of membership information secured individual participation in associational life without fear of recrimination. In *NAACP* v. *Alabama ex rel. Patterson* (1958), the Court found that Alabama could not require the National Association for the Advancement of Colored People (NAACP) to disclose its membership lists. The Court, comparing anonymous information giving to anonymous association, noted that "effective advocacy of both public and private points of view, particularly controversial ones, is undeniably enhanced by group association, as this Court has more than once recognized by remarking upon the close nexus between the freedoms of speech and assembly." More recently, in *McIntyre* v. *Ohio Election Commission* (1995), the Court struck down an Ohio law that required the disclosure of the name and address of the issuer of political literature on the literature. In its opinion, the Court stated that this requirement was a burden on political speech. Its argument was that anonymous leafleting was a continuation of centuries of protecting anonymous political discourse: "Under our Constitution, anonymous pamphleteering is not a pernicious, fraudulent practice, but an honorable tradition of advocacy and of dissent. Anonymity is a shield from the tyranny of the majority." The legal and philosophical ideas of anonymity and decisional autonomy to reveal or conceal, long observed in our political tradition, are central to the preservation of associational freedom. Yet claims to individual anonymity and decisional autonomy are not unfettered rights. In times of political conflict or strife when the paramount concern is national security or the protection of the common good, the diminution of these individual rights often occurs. For example, in 1951, the Supreme Court affirmed the conviction of eleven members of the Communist party under the Smith Act in *Dennis* v. *United States*. In *Dennis,* leaders of the Communist party were indicted in a federal district court under section 3 of the Smith Act for conspiracy to organize as the Communist party and advocate the overthrow and destruction of the U.S. government. Assessing the liability for dangerous speech was, in this case, based on an evaluation of the gravity of the potential harm, not the likelihood of the speakers' actually overthrowing the government. The immediate danger to the country was of paramount importance, relegating the interest of the individual to the protection of a larger sense of common good. The rhetoric of justification pointed to the responsibility of government to deal with dangers in a proactive manner. Justice Fred Vinson epitomized this sentiment in his opinion in *Dennis:*

> The mere fact that from the period 1945 to 1948 petitioners' activities did not result in an attempt to overthrow the Government by force and violence is of course no answer to the fact that there was a group that was ready to make the attempt. The formation by petitioners of such a highly organized conspiracy, with rigidly disciplined members subject to call when the leaders, these petitioners, felt that the time had come for action, coupled with the inflammable nature of world conditions, similar uprisings in other countries, and the touch-and-go nature of our relations with countries with whom petitioners were in the very least ideologically attuned, convince us that their convictions were justified on this score. . . . If the ingredients of the reaction are present, we cannot bind the Government to wait until the catalyst is added.

Vinson's argument reflects a rhetoric of justification that champions national security in a political crisis and potentially undermines individual claims to anonymity, decisional autonomy, and freedom of association. In *Dennis,* the mere advocacy of subversive activity justified governmental intervention. The use of investigative techniques to identify potential communists was a necessary tool to curb dissident activity before it resulted in the violent overthrow of the government. *Dennis* represents an example of a powerful and persuasive logic when threats to national security are on the public policy agenda.

The rhetoric of justification in *Dennis* was common to the investigative techniques in the red scare, civil rights era, during the Vietnam War, and the general rise of surveillance in the information age and in light of terrorism. During these points of historical conflict, defined by an external or internal threat, the diminution of individual rights of anonymity, decisional autonomy followed. The identification of an internal or external enemy of freedom and democracy was a powerful rhetorical device in justifying the use of surveillance, which sought to ensure security against communist infiltration—promoted either directly in the form of communist and socialist organizations or indirectly in the form of the civil rights movement and protests against the Vietnam War. In the case of the red scare, the communist threat represented an alternative political ideology that was perceived to be counter to the core of democratic values. The quest for the protection of democracy required the cleansing of the nation of the subversive element of communism in order to secure the future of freedom.

There are of course analogies that can be drawn to fighting terrorism and even to the public policy mandates of fighting crime or a war on drugs. For example, in the days following the September 11 attacks, President Bush contended that the United States "will rid the world of evil-doers." This is commanding rhetoric to justify the use of surveillance technologies as a means to counter these threats to society, identifying potentially criminal or terrorist activity before it occurs. Yet this rhetoric did not eliminate the concerns of anonymity and decisional autonomy; in fact, the values, norms, and expectations associated with anonymity and decisional autonomy figured prominently in the societal acceptance of biometric systems of identification.

The questions presented to focus groups and survey respondents in the research were designed to juxtapose considerations of anonymity and decisional autonomy against institutional objectives that sought to manage certain risks. In other words, the values, norms, and expectations of protecting anonymity and decisional autonomy were weighed against the

Table 4.1
Average ratings assigned by biometric users and nonusers: "How acceptable would it be to you if you or others were required to provide a biometric identifier in each of the following situations?"

Situation for use of biometrics	Biometric user mean (sd)	Biometric nonuser mean (sd)	Total mean (sd)
Boarding a plane	4.15 (1.16)	4.81 (0.40)	4.38 (1.01)
Accessing government buildings	4.44 (0.09)	4.29 (1.16)	4.38 (0.98)
Reentering the United States	4.18 (1.06)	4.58 (0.95)	4.33 (1.03)
Withdrawing money from a bank	3.88 (1.22)	3.97 (1.33)	3.91 (1.25)
Picking up a child from day care	3.42 (1.49)	3.70 (1.49)	3.52 (1.26)
Purchasing a gun	3.86 (1.35)	4.65 (0.55)	4.15 (1.18)
Making a credit card purchase	3.17 (1.39)	3.90 (1.42)	3.42 (1.43)
Included on your driver's license	3.21 (1.40)	3.58 (1.48)	3.35 (1.43)
Entering an event like a professional football game	2.10 (1.26)	2.77 (1.45)	2.28 (1.37)
Checking against terrorist watch list	3.85 (1.33)	4.45 (1.09)	4.05 (1.28)
Background check on foreigners	3.79 (1.28)	4.03 (1.38)	3.88 (1.31)
ID cards for guest workers	3.11 (1.43)	N/A	N/A
Group means	3.58 (1.21)	4.07 (1.15)	3.78 (1.23)

Note: Ratings were collected using a Likert scale from 1 to 5: 1 = Not at all acceptable, 2 = Somewhat unacceptable, 3 = Neither acceptable or not, 4 = Somewhat acceptable, and 5 = Very acceptable.

policy objective to be achieved by the institutions listed in table 4.1. The focus groups were asked to consider how acceptable it would be if they or others were required to provide a biometric identifier in different institutional contexts.

Some of the situational uses of biometric technology involved governmental objectives, such as border control, while other uses were designed to authenticate identity in transactions with financial institutions or to verify identity, as in a gun purchase. In each of the circumstances, the use of biometric technology implicated anonymity and decisional autonomy in a slightly different manner. Border control implicates the relationship between individuals and governmental objectives, while identity authentication in the case of financial transactions involves a consumer relationship with a private sector institution. In both cases, concerns with anonymity and decisional autonomy continue to prevail. These results were amplified in the national survey: respondents were asked to rank the acceptability of each situation in which biometric technology might be used on a scale of 1 to 5, with 1 representing "not at all acceptable" and 5 representing "very acceptable." The respondents' ratings indicated that checking against a terrorist watch group would be the most acceptable situation for biometric use, followed closely by reentering the United States and purchasing a gun. In contrast, their responses indicated that requiring biometric use in institutions outside the government were the least acceptable situations.

These ratings are very similar to those of the focus group participants; however, the survey respondents reportedly did not find biometrics as acceptable as the focus group participants did for accessing government buildings. Note in table 4.2 that the average "neutral" ratings of "when withdrawing money from a bank" and "when making a credit card purchase" also echo the focus group discussion results.

Table 4.2
Average reported levels of acceptability of various biometric situations (N = 1,000)

Situations	Mean (sd)
When boarding a plane	4.0 (1.4)
When accessing government buildings	3.9 (1.3)
When reentering the United States	4.2 (1.2)
When purchasing a gun	4.2 (1.3)
To check against a terrorist watch list	4.3 (1.2)
To do a background check on all foreigners	4.0 (1.4)
As part of identification for guest workers[a]	4.0 (1.2)
When withdrawing money from a bank	3.5 (1.5)
When making a credit card purchase	3.4 (1.5)
When entering an event like a professional football game	3.0 (1.5)
Instead of a debit card to make purchases at a participating store	3.2 (1.5)

[a]There was a statistically significant, negative correlation between age and reported level of acceptability of using biometrics as part of identification for guest workers ($r = -.11$, $p = .001$).

Participants across the focus groups and national survey tended to rate several uses of biometric technology as more acceptable than others: boarding a plane, accessing government buildings, checking against a terrorist watch list, and background checks on foreigners tended to be viewed as relatively more acceptable uses of biometric identifiers.

The individual who chooses to board a plane, reenter the country, or gain access to a governmental building exercises a decisional aspect of his or her associational life as it pertains to personal information. Physical access to a country, plane, or government building represented autonomous choices to reveal information and thus was not perceived to be a violation. Beyond this explanation, this reaction can also be explained by the ideological association of fighting terrorism with the use of biometric technology. According to the focus group participants, the acceptability of biometric identifiers in these circumstances reflects the close association made between the objectives of identification of terrorists after September 11 and biometric technology. Terrorism and the threat of future terrorist activity, for instance, prompted many respondents to report that the government could use biometric information in order to provide better security. This was a common theme reiterated in all of the focus groups. Most participants expressed concern about another terrorist attack and stated that the use of personal information to find potential terrorists was acceptable, and did not compromise autonomy because the pursuit of liberty was itself necessary to the preservation of liberty—for example, "I'm more secure about it. There's a lot of collection, not sharing [of your information], unless you're flagged as someone."

Security was also important to respondents in the national survey. When the goal was to prevent terrorist activity, the proposed use of biometric technology was considered more acceptable. For example, ratings were higher when the identity verification was against a terrorist watch list, a guest worker program, or prior to purchasing a gun. The rhetoric of a common good, such as public safety against terrorism, is a powerful justification for diminishing the importance of individual anonymity and limiting the exercise of decisional autonomy to a certain degree. This balance is not new. Historically the use of personal information to achieve social benefits or in pursuit of a common good has found policy justifications in testing newborns for HIV, revealing the identities of sex offenders, or balancing individual medical privacy with the war on drugs.[2] Amitai

Etzioni (1999).writes that "communitarianism holds that a good society seeks a carefully crafted balance between individual rights and social responsibilities, between liberty and the common good" (5).

The acceptance of biometric identification systems for fighting terrorism, however, is not without limits. Participants tended to feel that it was less acceptable for entering an event like a professional football game, stating that requiring biometric identification in this situation would be impractical and inconvenient. The wide-scale uses of biometric technology in public settings were viewed as a transgression of anonymity and decisional autonomy without sufficient policy justification or assurance that the identification could be carried out with accuracy and reliability. In these circumstances, anonymity and decisional autonomy took precedence over the potential benefit of identifying terrorists or criminals. When the inclusion of biometric information on drivers' licenses was considered, concerns also arose regarding violation of anonymity and decisional autonomy as a wider scope of information might be gleaned about their "private lives" in the large-scale deployment of biometric identification system. As individuals existing within society, we continue to remain autonomous by exercising some distinction between our public and private lives. Schoeman (1992) wrote:

> Clearly, people need some autonomy in order to engage in role-expressive aspects of self. The point of such autonomy, however, is not to disengage the person from the entire web of relations, but to enhance a feature of these relations, to make choices and counterbalances between relationships possible. Notions of privacy and intimacy are suggestive of these possibilities. (21)

The more practical matters of convenience were also important to participants. In contemplation of rising inaccuracies and inefficiency, social acceptability declines. For example, respondents in the national survey were asked to indicate how acceptable the use of biometrics to board a plane would be if one in five passengers was required to go through additional screening due to a false biometric reading. Once again, the respondents rated the acceptability on a scale of 1 to 5, with 1 representing "not at all acceptable" and 5 representing "very acceptable." With this new qualifying statement, the average level of acceptability dropped from 4.0 (above) to 3.7, with a standard deviation of 1.4. This finding is interesting for a variety of reasons. First, individuals were willing to allow a certain amount of inaccuracy when it came to strengthening security through the use of biometric technology. The potential social benefits of enhancing security with the use of biometric technology did not trigger an automatic perception that liberty was infringed. Yet there was a declining acceptability of the technology as the countervailing inconveniences rose. Perceptions of reliability and accuracy were thus important for the societal dimensions of acceptability beyond considerations of privacy.

Biometric identification also did not receive much support when it was used for the verification of identity of an individual picking up a child from day care. Only a minority reported that the use of biometric identification as a replacement for day care providers verifying identities would be "somewhat" or "very acceptable." One user suggested that the use of biometric technology at a child care facility diminishes the human factor of recognizing the right person to pick up a child and could prove to be more dangerous; that is, nothing short of knowing the person picking the child up would be secure enough for these participants. They were uncomfortable with technology usurping a primary relationship. The values, norms, and expectations associated with anonymity and decisional autonomy are also important for understanding this reaction. The revelation of private information when it involves relationships, as is the case with day care providers and children, is not maintained best through the medium of technology. The day care providers should be responsible for knowing which caregivers should pick up which children and those who should not. The widespread use of biometric identifiers to connect personal information, in the eyes of individuals, has the potential to transgress traditionally protected zones of decision making:

> In response to growing concerns about missing children, a state legislature decides to require all children attending private day care programs to be biometrically scanned for identification purposes. Parents object on the grounds that they are fully satisfied with the less-intrusive security already offered at the private day care facility and that biometric scanning will unduly traumatize their children. (Woodard, Orlans, and Higgens 2003, 162)

Disapproval of technology for some was extreme. Some focus group participants went so far as to suggest that there is no need for biometric technology and that it relies too much on technology and not enough on human interaction—for example, "Seems like they're trying to create an industry that's needless," and, "It's taking contact between people out of the picture. It's relying too much on machines."

The social descriptions of anonymity and decisional autonomy illustrated in the focus groups and national survey are instructive on several points. First, the use of biometric technology can potentially undermine anonymity and decisional autonomy by making an individual or his or her information visible when it is not perceived to be necessary. When respondents believed that the biometric identification of the individual in the circumstances of entering governmental buildings, crossing borders, or boarding an airplane administratively served a common good by virtue of a well-defined and limited policy objective and involved individual choice of physical access, the acceptance was broad.[3] Biometric systems of identification in situations necessitating greater security and, by extension of the logic, surveillance of physical activity were socially acceptable.[4] The use of biometric identification to protect transactions involving personal information was also socially acceptable. Across the groups, participants rated biometric identifiers as "somewhat" or "very acceptable" for withdrawing money from a bank or for identity verification in credit card purchases. In these circumstances, decisional autonomy was enhanced with the use of a biometric identifier by ensuring protection. In some cases, the use of a biometric identifier would be particularly appealing as a second level of security (e.g., in addition to a PIN). Many participants expressed concern over a PIN serving as the only identifier, suggesting that many people use the same PIN for all of their online passwords. The uses of biometric technology in situations not governed by choice, and therefore not protective of decisional autonomy, were suspect. Broad surveillance or inclusion of a biometric identifier on a national identification card was considered beyond a well-defined policy objective.

The Role of Surveillance Technologies as State Power

Perceptions of biometric technology point to the nuances of acceptability of the tools of surveillance technologies; anonymity and decisional autonomy serve to define the contours of this acceptance. The irony of surveillance technologies, illustrated well in the discussions of the focus groups and results of the national survey, is that for associational life in civil society to exist, there must be a sense of safety and confidence, which necessarily entails state power, even if it is in the form of technologies that regulate human behavior. Surveillance technologies are then, in light of the information age and September 11, part of modern state machinery necessary to the preservation of civil society; at the same time, they must abide by the limits imposed by anonymity and decisional autonomy.

Surveillance technologies conceived of as an extension of state authority require reconsideration of the notion of power. It is the disciplinary potential of surveillance that can fundamentally alter activities in civil society. Instead of the state exerting control after rules have been violated, the use of surveillance technologies attempts to manage the risk of crime or terrorism. Michael McCahill (1998) explains that this type of social control does not attempt to alter the individual offender but instead "alter(s) the physical and social structures in which individuals behave" (54). In this modern conception of state control, power is exercised by the acquisition and analysis of knowledge gathered through surveillance to manage risk of crime, fraud, and terrorism.

Yet as history reminds us, political action, especially for prospective danger, can devolve into a capricious political process that threatens the legitimacy of state authority. During the McCarthy era, for instance, allegations were rife with inaccuracies and misinformation, and the damage to both the legitimacy of the state and the associational life of civil society was apparent. Walzer (2003) argues, "No state can survive for long if it is wholly alienated from civil society. It cannot outlast its own coercive machinery" (317). This is not to say that state power does not have its place, but rather that the exercise of its power must not alienate its citizenry. In fact, in light of the prevalence of information technologies and personal information and the risks they pose, state action may be necessary to preserve the essence of civil society. Walzer (2003) calls civil society "a project of projects; it requires many organizing strategies and new forms of state action . . . and, above all, a new recognition (to paraphrase a famous sentence) that the good life is in the details" (321). The good life, especially in times of threat, does not require the end of state involvement or the termination of the use of surveillance technologies. This issue has been rendered irrelevant. The nexus between our public lives and surveillance technologies, whether in the form of commercial or governmental observation, is a reality. The question now is how to manage associational life in the purview of state action to maintain the good life of civil society.

The large-scale acquisition of knowledge regarding the day-to-day activities of individuals, even if the underlying goal of the policy was to prevent terrorism, was a fear expressed during the focus groups and national survey. This fear is not without historical referent.

Surveillance technologies and the information they acquire represent a different form of state power from that used in the red scare, evoking self-discipline and control through the power of technological observation, but the effects are similar and the fears comparable. In the era of the red scare, state power in the form of observation and information gathering of a different sort was formidable. The threat to associational life was similarly driven by the acquisition of knowledge, often gathered clandestinely. Those suspected of communist activities were under attack by state power in a wide variety of circumstances, sometimes with little or no justification. It is perhaps this experience with the red scare that cautions our use of biometric technology for the purposes of wide-scale observation of associational life and defines the legal limits of state authority relative to civil society.[5] In the midst of the red scare, the Supreme Court delineated when the means undertaken by the state became too overbearing on associational life. The limitation of a well-defined administrative purpose in the acquisition of information is an important principle of limiting state power over the good life.

In *Thomas* v. *Collins* (1945), the Court held that a Texas law requiring the disclosure of union members to the secretary of state before soliciting new members did not coincide with the rights of free speech and assembly. The Court tacitly recognized that a law revealing the members of the union might have a detrimental effect on associational life by undermining the ability of individuals to organize. The Court argued that "the First Amendment gives freedom of mind the same security as freedom of conscience." It then went on to say that to support legislative intrusion into the domain of labor unions was constitutional only if "grave and impending public danger requires this." Absent a closely linked justification between registration and the purpose of the state power, the requirement served only to threaten individuals who were organizing.

The principles of the *Thomas* decision in light of societal perceptions of biometric technology point to a relationship among anonymity, decisional autonomy, associational life, and surveillance technologies. General public use, such as in the case of a football game, or use that invades the sphere of personal interactions, such as picking up a child from day care, is viewed as an unnecessary encroachment on associational life without a specific public policy purpose that serves to limit the power of the state. Also important is to prevent the state from engaging in preemptive attacks on individuals, as was the case in *Thomas*. The requirement that Thomas first register with the secretary of state and secure an organizer's card before he was able to address the union meetings thwarted the freedoms of speech and assembly—cornerstones of associational life. The Court held that "as a matter of principle a requirement of registration in order to make a public speech would seem generally incompatible with an exercise of the rights of free speech and free assembly. Lawful public assemblies, involving no element of grave and immediate danger to an interest the state is entitled to protection." The Court indicates that knowledge of identity should not be a prerequisite to the freedoms of speech and assembly, especially in light of no immediacy of the danger that is the appropriate venue for state power. Similarly, knowledge of identity should not be a prerequisite for entering a football game.

Clear administrative purposes, like verifying identity before reentering the country or requiring a biometric identifier before boarding a plane, do not undermine activities of associational life in the same way that general surveillance of individuals watching a professional football game does. The important consideration in *Thomas* and for the focus group participants was the purpose for which identification was used and the setting in which it was deployed. If there is an element of immediate danger to the interest the state is entitled to protect, then there is a potential argument that overrides one of anonymity and decisional autonomy. In the case of *Thomas,* the state interest was to regulate labor unions in the name of the public interest. Such regulation, however, was to be aimed at fraud or other abuses, not at the restriction of First Amendment freedoms. When the purpose of identification drifts from its limited administrative purview, the societal acceptance of it dissipates, viewing the requirement above and beyond that which is necessary.

If surveillance technologies are perceived as being used to sanction associational life by limiting the exercise of political activities or ideas with preemptive surveillance without limit, then the legitimacy of governmental power begins to erode. In the national survey, individuals expressed concern that surveillance technologies and the gathering of

Table 4.3
Reported levels of concern with biometric use (N = 1,000)

Statements	Mean (sd)
That government agencies might use biometrics to track their activities[a]	3.2 (1.5)
That private industry might use biometrics to track their activities	3.5 (1.4)
That their biometric information might be stolen	3.5 (1.5)

[a]There was a statistically significant difference in respondents' level of concern with government agencies using biometrics to track their activities; Democrats were more concerned than Republicans (F(df = 4) = 4.09, p = .001; Democrats: mean = 3.50, sd = 1.43; Republicans: mean = 3.00, sd = 1.49).

personal information might be used to track their activities (table 4.3). Recall that the specter of constant observation was perceived as having the potential to undermine the freedom with which individuals pursue associational activities.

This concern is not without precedent. During the red scare, the gathering of personal information related to day-to-day activities by the state for the purposes of fighting communism triggered this same type of apprehension. For instance, in *United States* v. *Rumlely* (1953), Rumlely was secretary of an organization known as the Committee for Constitutional Government, which, among other things, engaged in the sale of books of a particular political leaning. His actions attracted the interest of the House Select Committee on Lobbying Activities, and they asked him to disclose the names of those who made bulk purchases of these books for further distribution. When Rumlely refused to divulge the information, he was convicted because of his refusal to give testimony or to produce relevant papers "upon any matter" under congressional inquiry under House Resolution 298. Under the resolution empowering it to function, the Committee was "authorized and directed to conduct a study and investigation of (1) all lobbying activities intended to influence, encourage, promote, or retard legislation; and (2) all activities of agencies of the Federal Government intended to influence, encourage, promote, or retard legislation."

According to this piece of legislation, the committee was enabled to conduct a study and investigation of all lobbying activities intended to influence, encourage, promote, or retard legislation and all activities of agencies of the federal government intended to influence, encourage, promote, or retard legislation. In its decision, the Court considered whether Rumlely had a right not to disclose the purchasers of his books under the First Amendment. The Court argued that he did:

> Once the government can demand of a publisher the names of the purchasers of his publications, the free press as we know it disappears. Then the specter of a government agent will look over the shoulder of everyone who reads. The purchase of a book or pamphlet today may result in a subpoena tomorrow. Fear of criticism goes with every person into the bookstall. The subtle, imponderable pressures of the orthodox lay hold. Some will fear to read what is unpopular what the powers-that-be dislike. When the light of publicity may reach any student, any teacher, inquiry will be discouraged. The books and pamphlets that are critical of the administration, that preach an unpopular policy in domestic or foreign affairs, that are in disrepute in the orthodox school of thought will be suspect and subject to investigation.

The *Rumlely* Court decision also reflects the sentiment articulated in the national survey: fear of being tracked or observed in one's day-to-day life serves to undermine associational life by devaluing the anonymity and decisional autonomy necessary to freedom of association. If there are no limits to the use of state power in the form of surveillance and the use of personal information, then individuals begin to fear the constancy of government in their life. Avoidance of the chilling effect, even before the consequence of retribution, is the core principle of First Amendment jurisprudence that necessitates protection. Real or imagined, the chilling effect is nonetheless the same. *Rumlely* also makes clear the dangers of political pressure, particularly in a time of political orthodoxy. The Court writes that danger to be avoided with First Amendment jurisprudence occurs when "the subtle, imponderable pressures of orthodox lay hold." The pressure of communist subversion was significant during the anticommunist movement, giving rise to a political orthodoxy that called for expunging all of its political proponents.

The era of the red scare provides a point of doctrinal reflection for our own search for security. Not only must the presence of surveillance technologies be restricted according to purpose and setting, but the information procured must not be put to the purpose of tracking everyday activities without cause or without procedural guarantees of limitation. This distinction that arises in the dynamics of societal acceptance is defined by the difference between surveillance of place and surveillance of mobilities. As the respondents intuitively understood, the use of information and judgments based on surveillance of place is more reliable and accurate than that based on mobilities.

Surveillance of Mobilities

Surveillance and interpretation of the data acquired from day-to-day activities, the part and parcel of associational life, require judgment about and interpretation of the activities. The narrative constructed by those who observe runs the risk of inaccuracy, and an inaccurate judgment might fundamentally undermine associational life by attributing meaning to activity that is unintended by the actor or, worse, perceived as criminal or terrorist, when it is quite innocent.

While the process of information gathering on its face appears impartial and neutral, the apparent impartiality may mask actual discrimination based on misjudgments and misinterpretation. It may not only be misinterpretation of the value of associations but of ideas. Not surprisingly, deterrence in this form is carried out by social pressures that are perhaps even more powerful when they reflect the sentiment of the majority, such as in preventing terrorism. The concept of the tyranny of the majority is pertinent. Unpopular views, particularly in times of political strife, are often chilled in the face of disclosure or, by extension, information gathering. It is necessary to remember that it is "rebel and heretic for whom, to a large degree, the First Amendment protections were forged" (Comment 1961, 1084). It is the anticipation of technological or human error in judging the acquired data that is potentially destructive to First Amendment freedoms rather than the technology itself. Seemingly innocuous personal information may be used to unintended effect if safeguards are not in place.

Surveillance of mobilities was also an issue that arose in the civil rights era. During the contentious political debate surrounding desegregation, the FBI made an organized effort to undermine the ability of political organization in support of the civil rights movement. This type of surveillance of mobilities, while different from the use of surveillance technologies, nonetheless provoked a similar fear in individuals and a similar danger to associational freedoms. Surveillance of mobilities in the information age and in light of September 11 has become more prevalent, and the interpretation of the data has become more pressing. The danger of course is allowing the judgment of data regarding mobilities to sort individuals in invidious and detrimental ways. The use of surveillance techniques during the civil rights movement is but one example of when judgments of activities were informed by a bias of observation.

In order to limit and combat the occurrence of civil unrest associated with protests, states often required that civil rights organizations produce membership lists. Such was the case in *NAACP* v. *Alabama* (1958), in which a contempt order was issued against the NAACP for refusing to produce membership lists in accordance with a court order. The Court was asked to consider whether the forced production of membership lists served a compelling state interest. Prominent in the case was the issue of First Amendment freedoms. The NAACP sought to protect the anonymity of its memberships by withholding the membership list from the state of Alabama. Alabama, for its part, contended that the exclusive purpose for the request of the list was to determine whether the NAACP was conducting intrastate business in violation of the Alabama foreign corporation registration statute; the membership lists, it said, were expected to help resolve this question. However, the state was not able to substantiate the relationship between the requirement of the membership lists and the expressed state interests, and therefore, the First Amendment interests of the members of the NAACP prevailed as a constitutional concern.

The Court argued that "this Court has recognized the vital relationship between freedom to associate and privacy in one's associations. We think that the production order, in the respects here drawn in question, must be regarded as entailing the likelihood of a substantial restraint upon the exercise by petitioner's members of their right to freedom of association." The NAACP, in establishing the claim of substantial restraint of their First Amendment freedoms, showed that on past occasions, revelation of the identity of its rank-and-file members had exposed these members to economic reprisal, loss of employment, threat of physical coercion, and other manifestations of public hostility. It was the fear of reprisal and exposure that substantiated the chilling effect on the NAACP. The fear, in other words, rather than the actual consequences, was enough for the Court to conclude that a constitutional violation had taken place. The Court argued that the regulation might induce members to withdraw from the NAACP and dissuade others from joining it because of fear of exposure of their beliefs shown through their associations and of the consequences of this exposure. In the domain of these indispensable liberties, whether of speech, press, or association, the decisions of this Court recognize that abridgment of such rights, even though unintended, may inevitably follow from varied forms of governmental action.

The surveillance of mobility affecting associational freedom in this case resulted from a request of membership lists. The same danger can occur because of surveillance of day-to-day activities, or mobilities, that are then subject to judgment. This judgment, in light of September 11, can potentially misinterpret activities as being terrorist in nature. Surveillance of mobility, more so than surveillance of place, may have serious consequences for associational freedom and the value of anonymity. In the case of the NAACP, the disclosure of membership lists and organizational activities would induce members to withdraw and dissuade others from joining. The existence of political dissension and political context is the bedrock of democracy, and it cannot be chilled by potential misuse of information gathering, as was the case in NAACP.

Heterotopias or Dystopia?

The mediation of social freedom, of which both the associational life and state authority are a part, requires not the end of technological interface with political and commercial entities but the institution of safeguards and limitations to balance one with the other. The historical referents discussed in this chapter serve as a cautionary tale that speaks to the use of surveillance and its potential detrimental effects on anonymity and associational life. The first cautionary tale is one of judgment. The interaction between technology and individuals is not only the observing and the observed. Behind the lens of observation is the element of human judgment, often affected by its biases and motivations that undermine the objectivity promised by technology. Technological observation does not eliminate the weight of judgment, which must be constrained by legal and policy guarantees to ensure a procedural recourse to erroneous judgments. The second cautionary tale is one of limiting the gaze of observation to specific populations and according to narrowly construed policy objectives. The panoptical effect is less when the gaze is segmented, differentiated, and acknowledged by the observed. The third cautionary tale is related to a distinction between place and mobilities. The good life is one that is in movement, not chilled by fear of observation. Avoiding a chilling effect requires an attending perspective that does not follow individuals through the many aspects of their associational life. These cautionary tales protect the heterotopias of Foucault from dissolving into the dystopia of a brave new world.

5 Trust and Confidence

Confidence is that feeling by which the mind embarks in great and honorable courses with sure hope and trust in itself.
—Marcus Tullius Cicero

All you need in this life is ignorance and confidence; then success is sure.
—Mark Twain, Letter to Mrs. Foote, December 2, 1887

Trust and Confidence in Institutions: Facilitating Social Capital

The interest in trust and confidence appears most notably in discussions of its role in civic engagement and societal functioning, and within organizational settings. Seminal works like those of Robert Putnam (1993), which explored the role of trust in civic engagement, and Francis Fukuyama (1995), which acknowledged the critical function of trust in societal functioning, have fueled an exploration of the role of trust within social systems. The issues of transaction costs, sociability among individuals, and deference to authority have defined the landscape of methodological inquiry (Levi and Stoker 2000). The benefits of trust as a social resource are obvious. As a factor in social organizations, trust can help to defray transaction costs, increase sociability among organizational individuals, and facilitate relationships with those in positions in authority (Cook, Hardin, and Levi 2005).

A discussion of trust and confidence in institutions is also important to understanding the role that personal identifiers, including biometric identification systems, can play in the reliance on a currency of personal information in public and private sector institutions. Not all institutions are viewed with the same level of trust and confidence when it comes to handling personal information. A lack of trust and confidence can affect an individual's willingness to trade personal information or, in the case of mandated use of personal information, affect his or her allegiance to an institution. My research has found that variances in trust and confidence in institutions are a function of how personal information is acquired, protected, and managed. This finding extends to biometric identification systems. The social acceptance of biometric identifiers includes an assessment of trust and confidence in the institutions that make use of the information and involves a consideration of the practices and purposes associated with the protection of the information. Generally, without any trust and confidence in institutions to protect personal information, there is a loss of social capital, which is increasingly built on the exchange of personal information.

Social capital refers to participation in social networks, which requires access, obligation, reciprocity, and trust (Coleman 1988, 1990). Social capital—social connections, networks, norms—is made up of the exchange of personal information. Without the attendant norms and trust, social capital, including social networks, economic growth, and the effectiveness of governments, among other things, declines (Halperin 2005). As personal information occupies a greater role in these networks of social capital, it is important to understand the values, norms, and expectations that sustain its use as a currency in transactions, networking, and communication, as well as to anticipate the sanctions, both technological and regulatory, that sustain it. This emerging form of social capital is not entirely different from its historical antecedents.

Although the terminology of social capital is recent, its etymology is linked to Alexis de Tocqueville's idea of the importance of social groups to democracy in America. Tocqueville (1835) described the importance of associational life in this way: "Feelings and ideas are renewed, the heart enlarged, and the understanding developed only by the reciprocal action of men upon one another" (515). He wrote that civic participation was an important facet of democracy in America in preserving the space for freedom that promoted civic engagement:

> Americans of all ages, all stations in life, and all types of dispositions are forever forming associations. These are not only commercial and industrial associations in which all take part, but others of a thousand different types—religious,

moral, serious, futile, very general and very limited, immensely large and very minute. . . . Nothing, in my view, deserves more attention than the intellectual and moral association in America. (517)

Emile Durkheim ([1893] 1964) observed later that not only was associational life good for individuals; it sustained a nation, thereby suggesting that a nation can be maintained only if secondary groups exist in the social space between the state and the individual. The meaning of social capital is built on more than twenty years of research, tracing a definition that varies from notions of human capital to physical capital. Social capital, though a catchphrase among academics and politicians, is defined differently depending on the perspective, intent, and context of the definition, often referring to community, civil society, or social fabric (Halperin 2005). Glen Loury (1977) was perhaps the first to use the idea of social capital to describe the resources of family relations that contribute to a child's cognitive development. In the social sciences, Pierre Bourdieu (1991) contributed to the idea by differentiating capital into three forms—economic, cultural, and social capital—while describing the relationships in this way:

The social field can be described as a multi-dimensional space of positions such that each actual position can be defined in terms of a multi-dimensional system of coordinates whose values correspond to the values of the first pertinent variables. Agents are thus distributed, in the first dimension, according to the overall volume of the capital they posses and, in the second dimension, according to the composition of their capital—in other words, according to the relative weight of the different kinds of capital in the total set of their assets. (231)

James Coleman (1990) built on the idea of social capital by considering the resources generated by the social structure or networks from which can be derived resources, including information, trust, and reciprocity. He defined social capital in this manner:

Social capital is defined by its function. It is not a single entity but a variety of different entities, with two elements in common: they all consist of some aspect of social structures, and they facilitate certain actions of actors – whether persons or corporate actors – within the structure. Like other forms of capital and human capital, social capital is not completely fungible but may be specific to certain activities. (303)

In an era when interpersonal relationships are not the foundation for coordinated action, the facilitation of social capital necessarily involves interactions among individuals in the medium of personal information and institutions by technological means. (Cook, Hardin, and Levi 2005) The value of organizational and network social capital to society offers an explanation for why individuals cooperate with public and private institutions even when there may be some risk to their values, norms, and expectations associated with liberty. The importance of an associational life is obvious even in contemplation of risk:

> First, the organizations and networks can be directly enabling. That is to say, they can be the vehicles we use to accomplish various purposes. . . . Second they can be an arena in which we develop particular forms of human capital. . . . The question for network and organizational capital is how they benefit us. Do they benefit us by what they can do, that is, what the network or organization can do? Or do they benefit us by what they enable us to do through the particular networks or organizations? (Hardin 2006, 81)

Participating in the public sphere in the information age requires the use of personal information by institutions and networks for a wide range of purposes, including security, communication, networking, and the facilitation of commercial or administrative transactions. Much of our social fabric is woven with the exchange of personal information. The act of cooperation with institutions, in other words, is much more than a decision about retaining individual rights to personal information; instead, the assessment may include consideration of the institution with which the information is traded, the purposes to which it is put, and the broader social capital to be gained. Participation in a networked world makes for a distinct form of civic engagement, and thus a different form of social capital that is facilitated by institutions. Personal information sustains an existing and emerging set of networks, invoking and creating norms, values, and expectations that require sanctions to protect and sustain social capital, either technological or policy orientated. Building trust and confidence in these institutions is a necessary step for securing social capital built on the currency of personal information.

Individuals who are faced with a decision to have trust and confidence in the use of their personal information cannot simply rely on the trustworthiness of another individual. Social capital is created by the institutions, networks, and organizations that facilitate cooperation in interactions, communications, and transactions characterized by the exchange of information with other members of the social network (Herreros 2004). In this sense, the existence of obligation and reciprocity of information serves as the basis of social capital. This meaning is different from the reciprocity and obligation that Putnam (2000) addresses:

> Members of a community that follows the principle of generalized reciprocity—raking your leaves before they blow into your neighbors' yard, lending a dime to a stranger for a parking meter, buying a round of drinks the week you earn overtime, keeping an eye on a friend's house, taking turns bringing snacks to Sunday school, caring for the child of the crack-head one flight down—find that their self-interest is served, just as Hume's farmers would both have been better off by sharing their labors. (135)

Today it is not the reciprocity and obligation of raking leaves that holds our social fabric together. Instead, it is the reciprocity and obligation associated with the use of personal information that is the stuff of modern notions of social capital. Traditional forms of social capital, which were sustained by geographically located communities and neighborhoods, represented an associational life created by proximity. Trust and confidence in the institutions of associational life were built on knowledge of one's neighbors, community organizations, and local institutions. However, social capital in a world that makes use of personal information transcends geographical limits and involves communications, transactions, and participation with individuals and public and private sector institutions that do not base trust and confidence on familiarity. Instead, trust and confidence in these institutions is influenced by how well they manage and protect personal information as currency for these interactions in cyber-civil society.

This means that trust and confidence in institutions figure into the idea of social capital because without it, or an adequate substitute for it, collective action in the public sphere cannot be facilitated. Thus, social capital cannot survive without institutional capital (Hardin 2006). Institutional capital, whether it be policy or law, enables individuals to "make the most of their social capital" by enforcing reciprocity and obligation Hardin (2006) explains that

> trust remains vital in interpersonal relations, but participation in functional systems like the economy or politics is no longer a matter of personal relations. It requires confidence, but not trust. The coherency of personal relationships is defined by reciprocal trust. This is an impossible expectation when extended to government. A citizen, more accurately described, has confidence in government. (88)

The conditions of trust and confidence in institutions are especially important for understanding the factors that give rise to the social assessment of biometric identifiers. A citizen must have trust and confidence in institutions and the way in which they acquire, use, and share personal information in order to accept biometric identification systems. The assessment of societal acceptance of biometric identifiers must be understood in the context of a desire or need to cooperate with institutions in obtaining and using personal information, but must also include a consideration of the conditions of protecting the norms, values, and expectations of liberty.

Biometric identification systems may also facilitate the creation of social capital by enhancing security in transactions in the public sphere in either the form of identity assurance in personal transactions or the prevention of national security threats, but the acceptance of this role is dependent on trust and confidence in institutions, which, while making use of personal information, should continue to gird the values that are important to society.

There are, then, many levels of analysis to explore. First, are individuals more willing to share personal information with some institutions and not others, and, if so, what are the norms, values, and expectations that they have? Second, what are the components of trust and confidence in institutions when handling personal information, and how are these conditions created? And to the important point that Cook, Hardin, and Levi made, what sustains confidence and cooperation when interpersonal trust cannot govern relationships? Finally, given these considerations, what are the consequences for biometric identification systems?

What This Study Is Not

This study is not an effort in understanding trust and confidence in institutions more generally. While criticism has been leveled at surveys that attempt to evaluate the role of trust institutionally because of an inaccurate broad-brush approach, the direction of this research is defined by a different aspect of institutional confidence. This research explores confidence and trust in institutions as they pertain to the individual's agreement or acquiescence to cooperate in the acquisition and use of personal information. As a point of departure, the relationships between the concepts of trust and confidence must be considered.

At the outset it is necessary to define the terms of inquiry. The nature of trust is itself complicated. In the literature on trust, various definitions exist. One definition, described as "encapsulated interest," is a notion of trust in a person who takes as part of his or her own interest the interest of others (Hardin 1999, 2005). This definition of trust is cognitive and represents a choice to trust based on sufficient knowledge to make a judgment about the motivations so that trustworthiness can be assessed:

> My trust turns, however, not directly on the Trusted's interests per se, but on whether my own interests are encapsulated in the interests of the Trusted, that is, on whether the Trusted counts my interests as partly his or her own interests just because they are my interests. Hence there is some risk that my interests will trump yours and that I will therefore not fulfill your trust in me; and your trust will be limited to the degree to which you think my encapsulation of your interests gives them enough weight to trump other interests I have. (Hardin 2006, 19)

Because the continuation of the relationship is important, there is an incentive to act in a trustworthy fashion. Reputation of an institution may figure into this idea of encapsulated interest. A good reputation, as an individual or an institution, can build confidence or create the basis for cooperative relationships. In place of reputation, institutional safeguards or legal guarantees may exist to shore up the protection of interest invested in the relationship to mitigate against risk, while building the foundation for reliance and confidence. This notion of trust "implies that many interactions in which there is successful coordination or cooperation do not actually involve trust . . . whereas many discussions of trust take cooperation to be virtually defining proof of trust" (Cook et al. 2005, 1).

Consider the information a credit card company uses to facilitate a transaction. Here the interest of the individual is encapsulated in that of the credit card company; however, the encapsulated interest that promotes trustworthy behavior in the credit card company might not always be a powerful incentive, and on another level, the individual will not know with certainty whether the credit card company is protecting his or her interest as encapsulated within its own. The literature divides between theories of trust that rest in a rational choice model and those that originate from a social perspective. The rational perspective assumes instrumentality in the interactions in which the individual engages to secure individual gain and exchange of resources (Stigler 1950). In this model, calculable risk is the focal point of determining when individuals will trust. At the other end of this spectrum, a social conception of trust is at work when understanding individual interactions with groups or institutions. Clearly, trust grounded in knowledge-based assessments cannot govern our interactions with public and private sector institutions because it is often impossible to know exactly how personal information is being used: "When we take a serious look at distrust and its role in our lives, we immediately realize that much of the hand-wringing about the lack or decline of trust in some societies is overwrought....We can get people to cooperate with us even without trust between them and us" (Cook et al. 2005, 6).

What is remarkable then is not that distrust is present in our current political context, but rather that cooperation continues to exist.[1] The focus here is on the role of trust in confidence in institutions, which, particularly now, tends to facilitate cooperation with institutions.

Modernity, Institutions, and Confidence

Modernization generally, and the rise of the information age specifically, has transformed the dimensions of trust. The challenge is developing trust in an abstract system. Anthony Giddens (1991) describes an abstract system as more than the tangible institution and the individuals associated with it; instead, he says, an institutional presence is defined by the regulatory patterns that underpin institutions and enable them to function smoothly and consistently. The rise of abstract systems in the modern age transforms the role of trust in institutions (Cook et al. 2005). Although trust can be understood as entrusting an agent, confidence points to the performance of the institution and is distinct in character from interpersonal relationships. The differences in the qualities of trust and confidence—interpersonal versus abstract institutions—also point to different forms of societal norms, values, expectations, and sanctions. For instance, Ericson and Haggerty (1997) suggest that an abstract system is unable to use individual moral agency as a basis of trust and so must implement some institutional assurance against risk. The management of risks and contingent events thus creates the conditions of confidence. Yet managing risks and contingencies is a complex process, and only one aspect of it is clearly legal enforcement. Fukuyama (1996) writes:

> People who do not trust one another will end up co-operating only under a system of formal rules and regulations, which have to be negotiated, agreed to, litigated and enforced, sometimes by coercive means. This legal apparatus, serving as a substitute for trust, entails what economists call "transaction costs." Widespread distrust in a society, in other words, imposes a kind of tax on all norms of economic activity, a tax that high trust societies do not have to pay. (26–27)

Theorists such as Fukuyama have argued correctly that law in the place of societal relationships results in "transaction costs." This is why it is important to understand that while the use of law is an option, the regulatory patterns to which Giddens refers are something more than pure and simple law. Especially in the instance of handling personal information, including biometric systems of identification, a comprehensive regulatory system is not in place. What constitutes the conditions of trust and confidence when the use of biometric identifiers in different institutional contexts is considered? This question has relevance for more than biometric identifiers. Emerging technologies generate discussion about values, norms, and expectations—privacy, anonymity, governmental intervention—that must be fit into the discussion of confidence and trust in institutions. These moral issues are personally important to individuals, and here trust and confidence can contribute significantly to cooperation with the institutions that make use of them. For this reason, the evaluation of trust and confidence in this study is concerned with understanding the value implications of institutions handling personal information and the performance implications, both of which contribute to cooperation (Siegrist, Earle, and Gutscher 2007). As Cook, Harden, and Levi (2005) point out, "Trust is important in many interpersonal contexts, but it cannot carry the weight of making complex societies function productively and effectively. For that, we require institutions that make it possible for us to exchange and engage in commerce and joint efforts of all kinds" (1). It is necessary to understand intangible and tangible factors in evaluating the potential risk to individual interests when the use of biometric information is at stake.

Confidence in Institutions as a Component of Trust

To provide a baseline for the importance of confidence in institutions entrusted with personal information, survey respondents nationwide and participants in the focus groups were asked to answer a series of questions regarding their confidence in various institutions. Through evaluation of confidence, we can evaluate the regulatory process, elements of a system, and the pursuit of some objective or desired state (Das and Teng 1998).

On average, the respondents in the national survey reported a higher level of comfort with institutions that provide more personalized service, for example, their employer, primary bank, or medical provider (table 5.1). A higher percentage of respondents also reported that they were "very confident" with this type of institution's information handling. Retailers and credit card companies scored the lowest average comfort ratings and the highest percentages of "not confident at all" responses. There were

Table 5.1
Reported levels of confidence in various institutions ($N = 1{,}000$)

Institutions	Mean (sd)	Very confident (%)	Not at all confident (%)
State and local authorities	3.0 (1.3)	14.4	17.7
Law enforcement agencies	3.1 (1.4)	19.2	16.5
Internal Revenue Service	3.3 (1.5)	24.6	17.2
Department of Homeland Security	3.1 (1.5)	20.9	19.0
State department or registry of motor vehicles	3.1 (1.4)	19.1	15.2
Their primary bank	3.6 (1.4)	33.2	12.3
Their employer	3.6 (1.5)	25.6	10.8
Credit card companies	2.9 (1.5)	19.4	23.2
Their retirement plan provider	3.5 (1.4)	26.8	12.5
Insurance companies	3.2 (1.4)	22.0	14.8
Credit reporting agencies (e.g., TransUnion or Equifax)	2.9 (1.5)	17.2	22.0
Their medical care provider or doctor	3.6 (1.4)	35.4	12.2
A hospital where they receive care	3.5 (1.4)	29.9	12.6
Nonprofit institutions	2.9 (1.4)	14.0	18.5
Retail stores	2.7 (1.4)	13.9	22.7
Internet retailers	2.6 (1.5)	13.8	27.5
Mail order companies	2.7 (1.4)	13.8	25.9

no statistically significant differences in responses according to age, gender, education, or political affiliation except for a statistically significant difference in respondents' level of confidence with the Department of Homeland Security: Republicans were more comfortable than those who indicated that they were independent ($F(df = 4) = 5.53$, $p < .0001$; Republicans: mean = 3.43, sd = 1.32; Independents: mean = 2.91, sd = 1.38).

The survey respondents' answers paralleled those of the focus group participants, although the survey respondents were reportedly more confident in their employers and retirement plan providers than the focus groups (table 5.2).

Confidence in institutions to handle personal information may be derived from a variety of factors. Individuals might perceive institutions as sharing or representing a set of values that they consider important. This might be particularly true with regard to the higher ratings given doctors, hospitals, banks, and even the Department of Homeland Security. The shared values with respect to personal information given to doctors, hospitals, or banks may point to the belief that the information is being used for a specific purpose related to the care of the individual. With regard to the Department of Homeland Security, for instance, the shared value may be fighting terrorism. Performance is another issue that builds confidence between society and individuals in the use of personal information as a currency. Institutions such as the Department of Motor Vehicles and the Internal Revenue Service have an established record of performance, an important predictor of societal confidence. If trust and confidence is a combination of values and performance, then what are the components of a lack of confidence?

Understanding Lack of Confidence

For the nationwide survey respondents, a lack of confidence, as identified by a rating of 1 or 2 for the institutions listed in tables 5.1 and 5.2 facilitated follow-up questions asking respondents to explain their lack of confidence. In the survey, 790 respondents were not confident with how their personal information is collected, stored, or used by one or more of the institutions above. Similar to the focus group participants, the respondents reported that they lacked confidence for the following reasons:

Table 5.2
Average ratings assigned by biometric users and nonusers: "How confident are you that each of the following institutions would keep any personal information they have about you safe?"

Institution	Biometric user mean (sd)	Biometric nonuser mean (sd)	Total mean (sd)
Department of Homeland Security	3.55 (1.34)	2.82 (1.47)	3.30 (1.42)
Internal Revenue Service	3.76 (1.30)	3.10 (1.35)	3.53 (1.35)
State department or registry of motor vehicles	3.18 (1.38)	2.43 (1.22)	2.92 (1.36)
Your primary bank	4.00 (1.14)	3.50 (1.07)	3.82 (1.14)
Your employer	3.61 (1.16)	N/A	N/A
One of your credit card companies	2.78 (1.25)	2.23 (1.18)	2.58 (1.25)
Your retirement plan provider	3.46 (1.10)	3.35 (1.16)	3.42 (1.12)
Your company personnel files	3.18 (1.28)	3.46 (1.29)	3.27 (1.28)
Your medical provider–doctor	3.84 (1.29)	3.90 (1.26)	3.86 (1.27)
Your medical provider–hospital	3.69 (1.23)	3.87 (1.26)	3.76 (1.24)
Retailers such as Macy's, Target	1.93 (0.98)	1.67 (0.71)	1.84 (0.90)
Online retailers such as Land's End, Amazon	1.98 (1.08)	2.03 (1.03)	2.00 (1.06)
Credit reporting agencies	2.00 (1.05)	2.07 (1.28)	2.02 (1.13)
Your school (only if you are a student)	3.40 (1.32)	N/A	N/A
Group means	**3.16**	**2.87**	**3.10**

Note: Ratings were collected using a Likert scale from 1 to 5: 1 = not at all confident, 2 = not completely confident, 3 = neutral, 4 = somewhat confident, and 5 = very confident.

- The institutions may not be able to keep their information safely stored and protected even if they want to (71.3 percent).
- The institutions may misuse their information for their own purposes (67.8 percent).
- The institutions may share personal information without permission (72 percent).
- The nongovernmental institutions may share their information with the government (51.5 percent).

- The institutions may use the information against them (47.5 percent).
- People working at the institutions may misuse the information without the institution's knowing about it (68.2 percent).

The respondents also cited additional reasons for their lack of confidence with the institutions' information handling:

- The agencies or government, or both, are not trustworthy and may be careless.
- Certain agencies sell contact information to solicitors or other organizations.
- The institutions may be broken into, or data may be stolen or hacked through their computer system.
- They are concerned about terrorism and "the general state of the country."
- Other people do not need to know their personal information.
- There is a lack of control over who has access to the information.
- They have been a victim of identity theft or another personal experience.
- The agencies may violate their constitutional rights.

Components of Confidence

For some institutions confidence was given more freely. Medical providers (both doctors and hospitals) and primary banks tended to be ranked the highest. In discussions, the focus group participants offered several reasons for this higher level of confidence. First, medical information was reported to be somewhat contained: doctors and hospitals were not perceived as part of an industry network in which sharing information provides an advantage to members of the network. As one focus group participant pointed out, employees of medical providers and financial institutions have no incentive to share personal information for gain. Overall, participants across groups tended to agree that doctors did not benefit from sharing information, and therefore they stated that their private information was more secure if stored by medical providers or financial institutions.

The organizations that tended to be ranked lowest across all groups were credit card companies and retailers (both physical stores and online retailers). Inappropriate information sharing as a factor was cited as the reason behind the low ratings that these institutions received. Participants commented that the volume of communication (e.g., catalogues, offers) that they receive from retailers and credit card companies also affected their confidence and served as evidence that their private information was being shared. In addition, participants across groups noted that after making a large purchase with one company, they receive catalogues and offers from other retailers.

Information Sharing, Confidence, and Regulation

The expressed concern with inappropriate information sharing led to a discussion of regulation, particularly in the cases of medical providers and financial institutions, the most trusted institutions. The regulatory frameworks of the Health Insurance Portability Accountability Act (HIPAA)[2] and the Gramm-Leach-Bliley Act[3] were viewed as increasing the level of confidence in these institutions and their employees. The focus group participants specifically pointed to the importance of legal sanctions in their determination of confidence. While Cook, Hardin, and Levi (2005) highlight the many shortcomings of legally backed enforcement, there may be a confidence-building aspect to the mere existence of legal protections. As the authors point out, legal sanctions can lag behind the types of technology used to gather personal identifying information, and the difficulty of assigning responsibility and causation in an organizational setting is problematic. They contend that while the assignment of responsibility works well in a criminal law setting, the organizational matrix is confusing and overlapping in an organizational setting where many individuals might be involved. Despite their arguments, that a general relationship existed between perceived legal protections and confidence level was borne out in the results of the nationwide survey.

Protections also benefited the perception of certain institutions and their handling of personal information. There was a positive relationship between respondents' perceptions of protection of their personal information by state and federal laws and how confident respondents reported they were with how their personal information was collected, stored, or used by various institutions. As respondents' perceived level of protection

Table 5.3
Correlations between respondents' perceived protection by state and federal laws and reported confidence with how personal information is collected, stored, and used by various institutions

Institution	Correlation coefficients
State and local authorities	$r = .272, p < .0001$
Law enforcement agencies	$r = .271, p < .0001$
Internal Revenue Service	$r = .236, p < .0001$
Office of Homeland Security	$r = .259, p < .0001$
State department or registry of motor vehicles	$r = .266, p < .0001$
Primary bank	$r = .234, p < .0001$
Employer	$r = .213, p < .0001$
Credit card companies	$r = .202, p < .0001$
Retirement plan provider	$r = .235, p < .0001$
Insurance companies	$r = .215, p < .0001$
Credit reporting agencies	$r = .211, p < .0001$
Medical provider/doctor	$r = .189, p < .0001$
Hospital where they receive care	$r = .211, p < .0001$
Nonprofit organization	$r = .150, p < .0001$
Retail stores	$r = .150, p < .0001$
Internet retailers	$r = .109, p = .001$
Mail order companies	$r = .126, p = .001$

increased, so did their confidence level with the institutions' handling of their information. Table 5.3 summarizes the relationships that were statistically significant.

Privacy notices, another potential form of regulation, received mixed responses. With regard to the use of privacy notices and the implications of self-regulation, some participants commented that privacy policy notifications from institutions were both reassuring and disturbing: although these notifications list measures taken to protect privacy, they also list parties with whom the institution may share information. The timeliness of action taken in light of a security breech was perhaps more important, as this respondent noted: "I know in terms of keeping my data safe, with hacking, my bank had some hacking thing happen to it and they were right on it (notified everybody)."

Regarding the preferred method of protecting against privacy violations, respondents were asked to indicate their level of agreement with

Table 5.4
Reported levels of agreement with privacy statements (N = 1,000)

Statements	Mean (sd)
The government should develop strong laws to protect the privacy of my information.	4.4 (1.1)
I decide when and where my personal information is shared.	4.5 (1.0)
Strong laws will keep my personal information private.	4.0 (1.3)
My personal information will never be completely private, no matter what I do or what the government does.	4.0 (1.3)

four privacy statements on a scale of 1 to 5, with 1 representing "strongly disagree" and 5 representing "strongly agree." As shown in table 5.4, the respondents agreed with all four statements regarding the privacy of their information. The various respondents favored some regulation but also acknowledge that perfect privacy is unattainable even with regulatory efforts.

Codes of Behavior

Individual commitment to codes of behavior within institutions is a concept related to confidence in those institutions. Focus group participants mentioned this professional norm especially as it applies to medical providers. Organizations such as the American Medical Association have contended that the moral commitment from its professionals has the effect of acting as an incentive and guiding force. "Professional codes of behavior were once conceived as regulating the one-on-one interactions between individual professionals and their individual clients or patients so as to make the professional reliable agents despite the fact that the clients or patients would typically be unable to judge the competence and commitment of professionals" (Cook et al., 110). Focus group participants identified the competence of medical providers and financial institutions and their employees as an important factor in adjudging confidence, particularly when it comes to handling personal information. Although commitment might not be easily evaluated, the enforcement of competence might be a substitute to alleviating concerns. This may explain why focus group participants indicated that confidence and trust were threatened by lack of control over those within the organization when it came to handling personal information. The relationship between trust and unbiased decision making was evidenced by a respondent's comment that one aspect of a loss of trust might occur if medical information might be used to discriminate.

Another general feature that reportedly affected participants' confidence was how many people were perceived to have access to the information. The state registry of motor vehicles, for example, was seen as a repository for records to which many people have access. Retailers were also seen as organizations whose records could be accessed by many people (e.g., in department stores with high employee turnover). Two comments are relevant here: "The institutions are comprised of the people who work there," and, "The person taking your information at Target is less qualified to handle that information than someone at Homeland Security." The handling of personal information, in other words, was not always controlled at the upper echelons of authority. Some participants suggested that it is not confidence toward the institution that should be measured, but rather confidence in the workers who are responsible for securing private information. This sentiment is consistent with the arguments made by scholars that perceptions of confidence in institutions are also a function of the reliability and competence of the individuals within institutions (Ayres and Braithwaite 1992).

"Competence here is not generalized but domain specific. For example, in tax collections, the payers must have good reason to believe that the extractions are going toward the purposes for which they are intended and not into the pockets of corrupt individuals" (Cook et al. 2005, 160).

In this respect, the evaluation of institutional reliability is not necessarily a matter of the individual employees but rather the institutional arrangements for "managing coercion and enforcement and upholding transparency, accountability, and fairness" (Cook et al. 2005, 161) . Similarly, Tyler and Degoey (1996) have shown that trust is influenced by the actions and approaches taken by authorities within institutions. This notion of trust is built on a relational model of procedural justice and suggests that people care about fair treatment by authorities. Tyler and DeGoey have demonstrated that "according to the social model of trust, an authority's intentions to maintain respectful relations in decision-making processes are central to trust. Attributions of positive intent lead group members to trust the authority and take the obligation to accept his or her decisions onto themselves" (340).

Instrumental models would predict that authorities are considered trustworthy when they provide favorable outcomes and when people believe they have been given some degree of control over the decision-making process. The relational model suggests that trustworthiness is primarily determined by neutral, unbiased decision making and by the degree to which individuals feel they are treated with respect. The central idea of the relational model is that people are concerned about the benevolence of the motives of the authorities: neutral and respectful treatment suggests that authorities have good intentions and will not violate a person's sense of dignity and self-worth (Tyler and DeGoey 1996).

One way to ensure neutral decision making and benevolence in an institution is to provide procedural protections that the individual can access. This point is particularly applicable to the indication given by the respondents in the national survey that procedural guarantees for biometric information were important to societal acceptance.

Familiarity, Trust, and Confidence

Trust at a fundamental level is a cognitive response by individuals. It is not surprising, then, that familiarity in interpersonal relationships gives rise to increases in trust. Familiarity is a source of knowledge about an individual that can serve as a basis for trust. But is the same true with regard to technology? Does familiarity with biometric technology create the basis for trust, or confidence in its capabilities, and at the same time reduce concerns associated with it? In the results of the focus groups and national survey, familiarity with the technology tended to increase acceptability. For example, in the focus groups, for those biometric technologies about which users knew little, they were reluctant to suggest that biometric technology was a safe technological measure. To test how information about the technology would affect perceptions, the focus group nonusers heard a short description of what is stored in the use of biometrics (an algorithm rather than an attempt to directly copy features of the identifier). Participants were told that the data stored would not be sufficient to recreate the identifier itself. With this additional information, views on safety tended to rise. After participants learned this information, their ratings changed:

- Ratings for the safety of voice recognition climbed.
- Voice recognition was rated as highly as they were iris or retinal scans and fingerprints.
- The safety ratings for fingerprints and face and hand recognition tended to go up overall.

However, there was little increase in the perceived safety of keystroke and signature recognition.

With this additional information, responses from the group of older nonusers with lower levels of technology experience more closely resembled those of the other groups than they did before. This group rated the safety of fingerprints more highly than before and evaluated voice recognition as even safer. But keystroke and gait recognition were still rated somewhat lower for this group (although the group's ratings overall remained higher than those in the other three groups).

Familiarity with biometric technology also influenced acceptability in the national survey. In the survey, 188 of the respondents reported that they had used biometrics. Table 5.5 identifies which types of biometrics respondents used.

Most biometric users reported being infrequent users of the technology: 53.2 percent reported using it only a couple of times per year, while only 15.4 percent of biometric users reported using it daily (table 5.6).

Regardless of which technology they had used, approximately 59 percent of biometric users considered themselves experienced, while 40.9

Table 5.5

Biometrics reportedly used by respondents (N = 188)

Type of biometric	Percentage of respondents (%)
Facial recognition	1.6
Hand geometry	6.9
Voice recognition	3.7
Fingerprint recognition	71.8
Keystroke recognition	1.1
Signature recognition	6.4
Iris or retinal recognition	4.8
Doesn't know which type was used	3.7

Note: Totals add up to greater than 100 percent because more than one answer was possible.

Table 5.6
Frequency of biometric use (N = 188)

Characteristic	Frequency (%)
A couple of times per year	53.2
A couple of times per month	14.9
Weekly	6.4
A couple of times per week	3.7
Daily	15.4
Don't know	6.4

percent reported that they were inexperienced. Furthermore, 66 percent of users were confident about the security of using the technology during their first use. Almost 90 percent of users reportedly felt as safe or safer about the technology than when they first signed up to use it, while 9 percent reportedly did not feel as safe as they did on first use. Almost 80 percent of biometric users were required to use the technology, while 15.9 percent signed up to use it. Of this 15.9 percent, 16.6 percent needed access to work, 30 percent needed access to a nonwork-related secure site, 6.6 percent used the technology to conduct a financial transaction, and another 6.6 percent wanted to use a retailer's convenience program. Other reasons for signing up for the technology included entry to a safe deposit box, computer fingerprint scanning, identification, personal protection, and a general interest in biometrics.

Interestingly, there were statistically significant differences between respondents who had previously used biometrics and those who had not in their level of acceptability to use biometrics in various situations. Respondents who had used biometrics in the past were more accepting of using them than respondents who had not used biometrics previously in several different contexts. A statistical difference was observed with respect to boarding a plane ($t(985) = -2.97$, $p < 0.0001$); Respondents who had used biometrics in the past were more accepting of using biometrics when boarding a plane (mean = 4.11, sd = 1.26) than respondents who had not used biometrics previously (mean = 3.99, sd = 1.36). (See figure 5.1.)

In queries about access to government buildings (figure 5.2), there were statistically significant differences between those who had used biometrics in the past and those who had not with respect to accessing government buildings ($t(991) = -2.27$, $p < 0.0001$). Respondents who had used

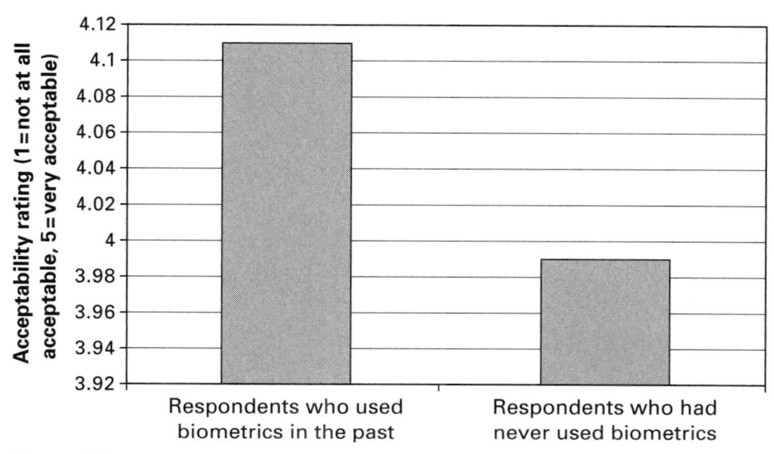

Figure 5.1
Average level of acceptability of using biometrics when boarding a plane

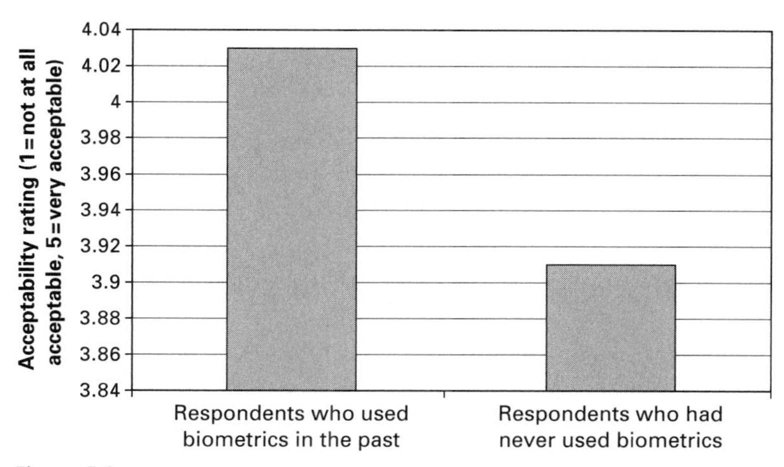

Figure 5.2
Average level of acceptability of using biometrics when accessing government buildings

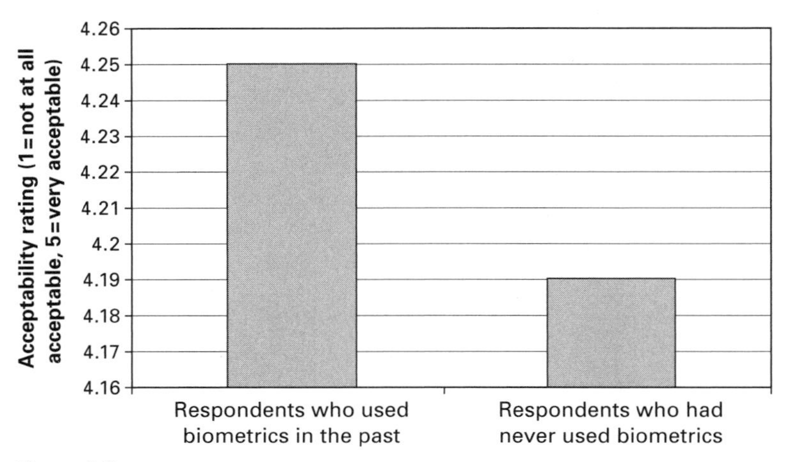

Figure 5.3
Average level of acceptability of using biometrics when purchasing a gun

biometrics in the past were more accepting of using biometrics when accessing government buildings (mean = 4.03, sd = 1.25) than those who had not used biometrics previously (mean = 3.91, sd = 1.34).

In addition, there was a statistically significant difference between the same groups with respect to purchasing a gun ($t(983) = -2.20$, $p < 0.0001$; figure 5.3). Respondents who had used biometrics in the past were more accepting of using biometrics when purchasing a gun (mean = 4.25, sd = 1.25) than those who had not used biometrics previously (mean = 4.19, sd = 1.35).

There was also a statistically significant difference with respect to withdrawing money from a bank ($t(988) = -1.054$, $p < 0.0001$; figure 5.4). Respondents who had used biometrics in the past were more accepting of using biometrics when withdrawing money from a bank (mean = 3.58, sd = 1.40) than those who had not used biometrics previously (mean = 3.55, sd = 1.46).

Beyond this direct exposure to biometric technology, 48 percent of respondents indicated that they were somewhat or very familiar with the technology through other sources:

- 40 percent had read about biometrics.
- 48.6 percent had seen something about biometrics on television or heard about it on the radio.
- 44 percent had seen a movie or read a book about the technology.

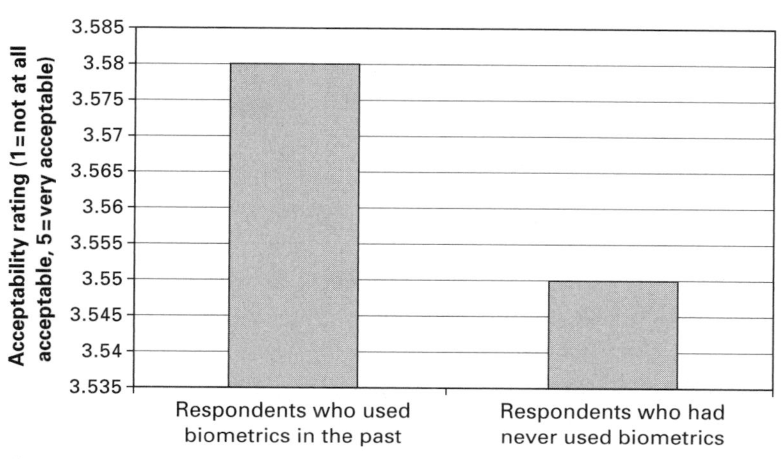

Figure 5.4
Average level of acceptability of using biometrics when withdrawing money from a bank

Interestingly, even indirect familiarity with biometric technology influenced the perception and, more important, acceptance. There was a statistically significant difference between respondents who had seen a movie about biometrics and those who had not seen a movie about it with respect to their perceived level of concern with identity theft ($t(690) = 3.41$, $p < 0.0001$). Respondents who had seen a movie about biometrics were less concerned with the technology (mean = 4.44, sd = 0.88) than those who had not seen a movie about biometrics (mean = 4.67, sd = 0.730). The accuracy of biometric technology in movies, books, and other sources could not be evaluated, but in some sense evaluation is irrelevant. These representations of biometric technology are part of the understanding of biometric technology, accurate or otherwise, and they are important to understanding the weight given to concerns raised in the proposed uses of the technology.

Although familiarity is an important basis for establishing interpersonal trust familiarity with respect to technology, it also reflects the human condition of developing horizons of understanding where knowledge is situational and dialogic (Gadamer 1997). The framing of biometric technology, whether in motion pictures, descriptions, or exposure to its use, influences acceptance of it. Not surprisingly, a lack of familiarity with technology translates into less societal acceptance. This insight is well represented in the literature on technology acceptance, where some of the most important considerations involve perceived usefulness of the technology; affective reaction toward using technology, and subjective norms of how the technology is important to the user and others (Taylor and Todd 1995, Venkatesh and Davis 2000).

This approach to user acceptance has generated a field of research designed to understand when users accept technology in a variety of contexts, including entertainment, organizational use, learning, or information seeking. These studies, usually conducted at the level of the user, indicate that the introduction of technology represents a process establishing usefulness, affective reaction, and a subjective norm of intended uses. In short, the introduction of a novel technology into society must build on familiarity and knowledge of it to establish the basis of trust and confidence in its proposed deployment, which includes considerations of institutional context.

Conclusion

Confidence in technology and institutions is determined by multiple factors. Confidence is, first, a matter of regulation or, more aptly put, perceived regulation, law and otherwise. Medical and financial institutions share the commonality of having a regulatory framework within which they operate and, effective or not, it tends to influence the levels of confidence that individuals afford to them. The presence of a regulatory framework supports the perception that individuals in the institution are answerable to a set of rules that carries with them consequences. Also important to confidence in institutions is the calculation of risk involved in sharing information. Individuals are willing to face some risk in the divulgence of personal information if the risk prevented is normatively higher. Whatever the degree of risk, individuals continue to look to institutions to provide fair treatment and neutral decision making. Individuals are willing to divulge personal information to achieve participation in the social realm of freedom, but they also expect a modicum of confidence in the institutions to which they provide information. Finally, confidence in technology is influenced greatly by exposure to and knowledge of it. Even indirect education about the workings of the technology increased social acceptance of it. Increasing confidence in the information technology and personal information in our transactions, interactions, and communications is important for society in an increasingly networked world.

Trust and confidence are said to play an important role in creating and maintaining the social capital of our democracy by facilitating civic participation and engagement. According to notable theorists such as Robert Putnam (2000), in the absence of trust, civic participation in networks and organizations tends to decline as does our engagement with government:

> People who trust others are all-round good citizens, and those more engaged in community life are less likely (even in private) to condone cheating on taxes, insurance claims, bank loan forms, and employment applications. Conversely, experimental psychologists have shown that people who believe that others are honest are themselves less likely to lie, cheat, or steal and are more likely to respect the rights of others. In that sense, civic engagement and social trust are mutually reinforcing. . . . Trust in government may be a cause of consequence of social trust, but it is not the same thing as social trust. (137)

The concern is that rising distrust and declining social participation characterize the state of civic engagement in America today unlike any other previous era. Putnam (2000) writes about a fundamental change between the World War II generation and the decline of public life in later decades:

> It wasn't so much that old members dropped out—at least not any more rapidly than age and the accidents of life had always meant. But community organizations were no longer continuously revitalized, as they had been in the past, by fresh sets of new members. Organizational leaders were flummoxed. For years they assumed their problem must have local roots or at least that it was peculiar to their organization, so they commissioned dozens of studies to recommend reforms. The slowdown was puzzling because for as long as anyone could remember, membership rolls and activity lists had lengthened steadily. (16)

If Putnam is correct, declining civic engagement and participation poses many costs. From Putnam's view, the less that society is engaged in common purposes in groups and organizations, the more likely the tendency is to have less trust and confidence in societal institutions. The causal direction is drawn from a supposed decline in civic engagement to a loss of civic participation and involvement with government.

Many things have altered or diminished the role of trust in society, as Cook and Hardin (2001) noted: "The long-term changes from small communities to mass urban complexes, mere coordination and state regulation have become far more important . . . while the actual role of trusting relations has declined relatively" (xx). Not only has the rise of mass urban complexes prompted interactions without the necessary safety net of trust, but the rise of divulgence of personal information has also taken place without the interpersonal context of trust. The events of September 11 also serve to heighten societal concerns of trust in public and private sector institutions.

The question of distrust or trust post–September 11 particularly affects societal perceptions of biometric identifiers and the institutions that make use of them. For instance, following September 11, biometric identifiers were touted as a form of personal identification that might ensure greater security. Whether it was their use in border security or proposals for passenger screening programs, biometric identifiers and other forms of personal information became a more important factor in the landscape of trust and confidence as institutions began to gather information for the purpose of confronting terrorism. The use of personal information, of course, has been on the rise in the age of technology and Internet transactions, but the events of September 11 ushered in the rhetoric of necessity and security. Personal information was required by governmental agencies to verify identities and ferret out the undercurrents of terrorism. At the same time, trust in these policy objectives was, in certain circumstances, called into question as the comingling of private and public sector information became apparent. Private organizations such as AT& T were found to be data mining at the behest of the National Security Agency for the purposes of undermining terrorist activity, which raised questions of trust and confidence for both private and governmental institutions.

Putnam (2002), however, argued that September 11 represented an opportunity for enhancing social capital. The potential for turning around the trend of increased loss of social capital and loss of trust in modern institutions was one that Putnam suggested might be changed by a crisis. In *Bowling Alone,* for instance, he had contended that the lack of "civic engagement in America would be eased by a palpable national crisis, like war or depression or natural disaster, but for better and for worse, America at the dawn of the new century faces no such galvanizing crisis." When September 11 occurred, Putnam wondered in *Bowling Together* (2002) if the event signaled a start to a new era or if it was, alternatively "a brief pause during which we looked up for a moment and then returned to our solitary pursuits."[4] During October and November 2001, Putnam and his colleagues returned to the original five hundred respondents of his survey conducted in 2000 that substantiated his work in *Bowling Alone* and tried to assess points of change and stability. He found a marked change in the usual societal malaise and suggested that

> trust in government, trust in the police, and interest in politics are all up. . . . This is no doubt partly the result of a spurt of patriotism and "rally round the flag" sentiment, but it also reflects a sharper appreciation of public institutions' role in addressing not just terrorism but other urgent national issues. The result? A dramatic and probably unprecedented burst of enthusiasm for the federal government. (2002)

Putnam's surveys showed that in the months following the attacks, a net change in trust in "the people running your country" grew by 19 percent in younger people. He also found that in answer to a standard question ("How much can you trust the government in Washington to do what is right: all of the time, most of the time, some of the time, or none of the time?") 51 percent of the respondents expressed greater confidence in the federal government in 2001 than a year earlier. Putnam argued that this trust in government translated into a possibility of civic renewal and social engagement and saw the opportunity presented by September 11 to be like that of World War II.[5] According to Putnam, "Americans who came of age just before and during World War II were enduringly molded by that crisis. All their lives, these Americans have voted more, joined more, given more. But the so-called Greatest Generation forged not merely moods and symbols, as important as those were; it also produced great national policies and institutions (such as the GI Bill) and community-minded personal practices (such as scrap drives and victory gardens)" (2002).

Mobilization can spur more subtle but profound changes. For instance, Putnam contends that the rhetoric of mobilization in World War II gave rise to changes in our own social fabric, including the salability of the civil rights movement. The sentiment captured immediately after September 11, he said, posed a similar opportunity.[6] So great was this potential that the Bush administration consulted Putnam as to how to capitalize on the opportunity. September 11 offered a chance to increase trust in government, but with the increased presence of the tools of information technology in the hands of the public and private sectors, there was also fear of abuse. Indeed, some critics were quick to claim that the American public accepts information technologies, including biometric systems of identification, because of confusion and fear; "the interest in surveillance cameras, data-profiling systems at airports, integrated databases of personal information, and biometric identification systems is one sign of our fears and confusion about whom to trust" (Rosen 2005, 8).

However, despite the claim of some that the American public was blindly following the integration of biometric technology proposed because of the events of September 11, users and potential users of biometric technology have assessed biometrics based on a multitude of considerations rather than adopting the beliefs and values handed to them, whether from the perspective of the advocates or the critics. The choices and judgments about the integration of biometric technology into our lives is not linear or simplistically a function of a crowd mentality. Instead, Americans are faced with the simultaneous challenge of reevaluating the organizations and institutions on which they rely for their security while dealing with the new reality that security and well-being require the use of greater amounts of personal information. Ultimately this requires boosting the trust and confidence in the public and private institutions that facilitate the use of personal information, which can include a role for biometric identifiers.

6 Paternalism

If men were angels, no government would be necessary. If angels were to govern men, neither external nor internal controls on government would be necessary. In framing a government which is to be administered by men over men, the great difficulty lies in this: you must first enable the government to control the governed; and in the next place oblige it to control itself.
—James Madison, *Federalist Papers*, No. 51, 1787

If I knew for a certainty that a man was coming to my house with the conscious design of doing me good, I should run for my life.
—Henry David Thoreau, *Walden*, 1912

Paternalism and Its Limits

In the information society, there is an ever-growing need to authenticate individuals. Biometric technology has emerged as a means for authentication, but it has still to fully win over societal acceptance. Societal apprehensions regarding biometric technologies include the encroachment on individual liberty by public and private sector institutions, which rely on identity assurance for objectives ranging from combating consumer fraud to fighting the war on terror. These regulating technologies—or those that regulate or influence human behavior—are novel in many ways, but they still represent a form of control over individuals, an issue long debated in discussions over the appropriate balance between authority and the preservation of the liberty of the individual, which takes place under the philosophical auspices of paternalism.

It is not only in the realm of crime and law enforcement that technologies can act to regulate behavior. Biometric identification systems are part of the spectrum of regulating technologies and are now part and parcel of our public and private sectors. Biometric technologies are used by private sector companies to combat fraud with identity assurance to regulate consumer behavior and by the public sector to pursue state objectives, such as fighting terrorism. However, while information and intelligence have long served the purpose of regulating behavior, biometric technologies may change the usual means of protecting individual liberty. "The idea that our bodies can be reduced to a means by the state—that the human body itself can be a crime control technology—offends human rights at its very roots in human dignity" (Bowling, Marks, and Murphy 2008, 75). The potential benefit of biometric identification systems that regulate human behavior—combating identity theft, financial fraud, and terrorist activity, among other things—must continue to abide by the guarantees of individual liberty. Striking this balance raises a perennial question of balancing the use of authority to combat harm and protect the common good while sustaining individual liberty. The philosophical idea of paternalism provides a framework for evaluating the regulative and constitutive aspects of biometric technology while preserving individual liberty in a wired world.

Paternalism has a long history in the American philosophical tradition, representing a deeply rooted tension about how to best preserve the liberty of the individual against encroachment by the state. Familiar forms of paternalism include welfare programs, mandatory vaccination requirements, and legal restrictions on certain drugs by the Food and Drug Administration.[1] A debate over the need to regulate the influence of the Internet introduced paternalism to the information age. Cyberpaternalists argued that the lack of traditional regulatory control systems did not mean total freedom within cyberspace (Lessig 2000). To the contrary, they contended that regulation of human behavior was being driven by the market interests of those who occupy the cyber-commons—by the design of the technology and by actors who implement it, affecting human behavior with what Lawrence Lessig (2000) has described as "code," or the hardware and software that regulate human behavior explicitly or implicitly. Lessig argued that that liberty that might have been in place during the founding of the Internet was increasingly threatened by a gradual loss of values considered fundamental to liberty. As an analogy, he used the values manifested in our constitutional founding, where structural and substantive considerations went hand in hand to preserve the values that informed the Constitution. From Lessig's perspective, the development of the Internet poses the same challenge as the Constitution. As the structure of the Internet is developed, it is necessary to be cognizant of the substantive values and norms of liberty in the structure of technology, or they will be lost.

The concern over the regulation of the Internet has evolved into a larger discussion about the development and implementation of information technology so that the regulating influence of information technology on human behavior accords with our societal norms, values, and expectations.

The Place of Paternalism

One of most fundamental substantive values at stake in the debate about biometric systems of identification is the idea that individual liberty should be protected from encroachment even when the action taken is designed to protect the interest of the common good or to guard against harm to society at large. Actions taken in defense against harms or in pursuit of the interest of the common good raise a long-standing problem for individual liberty. As both Madison and Thoreau observed, the problem is the proper balancing of paternalistic intervention against the liberty of the governed. Encroachment on individual liberty, often justified in terms of the welfare, needs, interests, or good of the persons subject to it, is often met with skepticism and resistance. The same dilemma occurs with regard to the proposed use of technologies, such as biometric systems of identification, to regulate behavior. The idea of paternalism can help to identify the tensions between the need to protect against harm and pursue societal interests while protecting individual liberty; more important, paternalism can aid in understanding why and when society accepts paternalistic intervention.

To fully comprehend the philosophical place of paternalism in our liberal political discourse would be impossible without taking account of John Stuart Mill's insights.[2] It may be an understatement to say that Mill leaves us with a healthy suspicion of paternalism. He viewed governmental intervention as a threat to individual liberty, but acknowledged that there was an inevitable place for it:

The maxims are, first, that the individual is not accountable to society for his actions in so far as these concern no one but himself. Advice, instruction, persuasion and avoidance by other people, if thought necessary by them for their own good, are the only measures by which society can justifiably express its dislike or disapprobation of his conduct. Secondly, that for such actions as are prejudicial to the interests of others, the individual is accountable and may be subjected either to social or to legal punishment if society is of the opinion that the one or the other is requisite for its protection. (Mill [1859] 1956, 14)

The justification of paternalistic intervention is juxtaposed against the idea that individuals are best able to protect their liberty because they possess the particularized knowledge to pursue the best course of action as it comports with their self-interest. Mill asserted that "with respect to his own feelings and circumstances, the most ordinary man or woman has means of knowledge immeasurably surpassing those that can be possessed by anyone else" (Mill 1869, 207). In the philosophical context of liberalism, criticism of paternalistic intervention is based on the idea that it undermines the exercise of free choice as an expression of liberty because the state cannot know as well as the individual the best course of action. As Julie White (2000) summarized, "Mill seems to criticize intervention for what I will argue are two distinct—that is, separable—reasons: both because it violates individual choice as the expression value of liberty, and because the knowledge in which choice is based is highly particular, and therefore intervention would likely be misapplied" (129).

Despite these general prohibitions, Mill occasionally acknowledges that paternalistic intervention is appropriate. According to Mill, if an individual is making a misinformed choice that undermines future liberty, then paternalism may be appropriate. For instance, if an individual chooses to sell himself into slavery, this action cannot be construed as a well-informed choice because, from Mill's perspective, it precludes future choices and thus freedom. The justification of paternalism rests in the assertion that in order to continue to be free, individuals must be prevented from abdicating their own freedom:

The ground for thus limiting his power of voluntarily disposing of his own lot in life is apparent, and is very clearly seen in this extreme case. The reason for not interfering, unless for the sake of others, with a person's voluntary acts is consideration for his liberty. His voluntary choice is evidence that what he chooses is desirable, or at least endurable, to him, and his good is on the whole best provided for by allowing him to take his own means of pursuing it. But by selling himself for a slave, he abdicates his liberty; he foregoes any future use of it beyond the single act. He therefore defeats, in his own case, himself. He is no longer free; but is thenceforth in a position which has no longer the presumption in its favor, that would be afforded by his voluntarily remaining in it. The principle of freedom cannot require that he should be free not to be free. (Mill 1869, xx)

Mill also contended that if an individual is taking action that indicates that he or she is ignorant of the consequences and is therefore not acting in what might logically be construed as self-interest, then paternalistic intervention may also be appropriate: "If a public officer or anyone else saw a person attempting to cross a bridge which had been ascertained to be unsafe, and there were not time to warn him of his danger, they might seize him and turn him back, without any real infringement of his liberty; for liberty consists in doing what one desires, and he does not desire to fall into the river" (Mill 1869, 117). In this example, the individual is not acting in a way that could be construed as consistent with self-interest, making paternalistic intervention relatively unproblematic. The assumption is that if the individual acts with a certain degree of ignorance or incompetence, then paternalistic intervention is justified. From Mill's perspective, preventing an individual from falling off a bridge is not in conflict with the individual's freedom of choice because no individual would likely choose to fall off a bridge. The legitimate basis of paternalism in this instance is the ignorance of the harm. In Mill's hypothetical of slavery, the individual is precluding future liberty with his or her choice, and this justified paternalistic intervention. In the case of falling off the bridge, there is a harm to be prevented because of ignorance or incompetence on the part of the individual. In both of these examples—falling off a bridge or selling oneself into slavery—the individual must be protected against the wrongful invasion of interest with paternalistic intervention.

The harms to be avoided in the cases of falling off a bridge and selling oneself into slavery, however, are different from justifications of paternalistic intervention that are based on normative judgments about what is best for the individual. When paternalistic intervention is justified in terms of increasing an individual's happiness or compelling a wiser choice, Mill believed that the liberty of the individual is threatened:

That the only purpose for which power can be rightfully exercised over any member of a civilized community, against his will, is to prevent harm to others. His own good, either physical or moral, is not a sufficient warrant. He cannot be rightfully compelled to do or forbear because it will be better for him to do so, because it will make him happier, because, in the opinion of others, to do so would be wise or even right. Those are good reasons for remonstrating with him, but not for compelling him or visiting him with any evil in case he does otherwise. (Mill, 1869, 7)

Normative justifications that rest on the good of the collective or to promote the good of the individual must always be weighed against what Mill viewed as the greater evil: the domination of the individual's liberty. Even the achievement of a good or the prevention of harm interferes with the liberal commitment to allow for the liberty of self-determination, as Mill explained:

But neither one person, nor any number of persons is warranted in saying to another human creature of ripe years that he shall not do with his life for his own benefit what he chooses to do with it. He is the person most interested in his own well being . . . with respect to his own feelings and circumstances, the most ordinary man or woman has means of knowledge immeasurably surpassing those that can be possessed by anyone else. The interference of society to overrule his judgment and purposes in what only regards himself must be grounded on general presumptions which may be altogether wrong, and even if right, are as likely as not to be misapplied to individual cases, by persons no better acquainted with the circumstances of such cases than those are who look at them merely from without. ([1859] 1956, 93)

Here, Mill illustrates an assumed tension in liberal political culture between paternalism and individual liberty, which may not be apt as it is applied to information technologies and the information they generate. There is a tendency to assume that the individual is always in the best position to act in his or her best interest; the components of knowledge and choice are thus assumed. But as White (2000) demonstrated, justification for paternalistic intervention is unnecessarily limited because of this bias:

In our own political practices, we lean heavily to interpreting most action of individuals as self-regarding, and the tension between self-and other-regarding has largely been dealt with through the establishment of a space for, and limits to, individual rights. It is in the context of these commitments and against the backdrop of liberal political culture that paternalism becomes such a forceful criticism. For paternalism names the failure to respect the capacity and right of self-regarding individuals to define his or her own good. (13)

In the case of distribution of welfare benefits, for example, the emphasis on individuals as self-regarding results in negative connotations for those who must make use of these benefits. Individuals on welfare are described as "dependent" and "needy," eliciting the rhetoric of a flawed individual (Fraser and Gordon 1994). The end result is a judgment leveled against paternalistic intervention based on the venerable principles of self-sufficiency and autonomy. As Judith Shklar (1990) observed, it is the assumption of incompetence that makes paternalistic intervention distasteful in our political tradition:

> Paternalism is usually faulted for limiting our freedom by forcing us to act for our own good. It is also, and possibly more significantly, unjust and bound to arouse a sense of injustice. Paternalistic laws may have as much consent as any other, but what makes their implementation objectionable is the refusal to explain to their purported beneficiaries why they must alter their conduct or comply with protective regulations. People are assumed to be incompetent without any proof. (119)

This insight can help explain the mistaken assumption that technologies that prevent harm to individuals translate neatly into a violation of individual liberty. In fact, the individual can know neither the range of future choices nor the harms to be avoided, making paternalistic intervention an attractive option that has the effect of preserving the common good and individual liberty because it enhances the currency of personal information by protecting against threats even though individual choice is overridden.

An example is illustrative. Microsoft and Google have begun offering Web-based personal health records. This system is described as creating a health information economy in which consumers play a role in deciding when and to whom to release their records. Although this form of individual control has salience in terms of preserving individual choice and expression of self-interest, the technology leaves unanswered the question of whether there is sufficient knowledge for an individual to make this choice competently. Microsoft and Google are not subject to the restrictions of the Health Insurance Portability and Accountability Act. The lack of paternalistic intervention means that unregulated Web systems are vulnerable to marketing and manipulation of patient information by advertisers and others. Dr. Isaac S. Kohane said, "I'm a great believer in patient autonomy in general, but there is going to have to be some measure of limited paternalism" (Lohr 2008. In this sense, paternalistic intervention may be an essential tool of enhancing the currency of personal information by allowing some regulating influence of information technology or policy. Under certain circumstances, the preservation of autonomy involves paternalistic intervention because individual choice, which requires sufficient knowledge to exercise it, does not exist. Paternalistic intervention in these instances ensures competency rather than demonstrates incompetency. Vandeveer (1968) explained,

For example, whether a child can play safely in her yard may depend in part on whether an expressway has been built adjacent to it and whether or not barriers between the yard and the expressway have been erected to render the yard a safe area. If there is no barrier and the child cannot play safely in the yard, we might rightly say that the child is not, in the sense previously defined, competent to play in the yard. (358)

Similarly, the risks of identity theft or the threat of another terrorist attack cannot be managed well individually because the particularized knowledge that is necessary is lacking to guard against them. Background checks on gun purchasers, verification of foreign workers, or the use of terrorist watch lists are all regulatory effects that diminish risks that arise and which an individual is not competent to manage alone. Individuals, in assessing the acceptability of paternalism are very much aware of the risks that threaten data security and cognizant that something beyond their control is necessary to protect against them. Recall the perceptions of risks associated with their data security in the national survey (see table 6.1).

Table 6.1
Concerns of data security

Issues	Mean (sd)
Unwanted telemarketing	4.2 (1.2)
Unwanted e-mails or spam	4.3 (1.2)
Another terrorist attack within the United States	4.2 (1.8)
Identity theft	4.5 (0.9)
Someone revealing your personal information to government without your permission	4.3 (1.1)
Someone revealing your personal information to private companies without your permission	4.6 (0.8)
Someone getting into your bank accounts without your permission	4.6 (0.9)
Loss of our civil liberties as part of the war on terror	4.1 (1.3)
Illegal aliens in the United States	3.8 (1.4)

The perception and understanding of these risks to data security has influenced the acceptance of biometric identifiers. As a form of identity assurance, biometric identifiers are part of paternalistic protection that defends against the type of risk feared. Consider that for those who feared the possibility of another terrorist attack, the acceptance of biometric systems of identification as a form of identity assurance was more likely in a variety of circumstances. As the respondents' level of concern with another terrorist attack increased, so did their level of acceptance of using biometrics. This correlation was also present with other perceptions of risks or harms to be avoided due to manipulation of identity, including accessing a governmental building, purchasing a gun, or checking against a terrorist watch list. The perception of risk also influenced the tolerance for inconvenience. Individuals who perceived the possibility of another terrorist attack to be likely were also more accepting of the possible inconvenience of misidentification. Table 6.2 outlines the statistically significant relationships.

There is a clear relationship between a perception of risk and the acceptability of biometric identifiers. When an individual is unable to control for risks, the acceptance of paternalistic intervention, including

Table 6.2

Correlations between respondents' reported level of concern of threats and their perceived acceptability of using biometrics in various situations

Biometric uses	Correlation coefficients
When boarding a plane	$r = .145, p < .0001$
When accessing government buildings	$r = .166, p < .0001$
When reentering the United States	$r = .150, p < .0001$
When purchasing a gun	$r = .119, p = .0002$
To check against a terrorist watch list	$r = .220, p < .0001$
To do a background check on all foreigners	$r = .278, p < .0001$
As part of identification for guest workers	$r = .232, p < .0001$
When withdrawing money from a bank	$r = .134, p < .0001$
When making a credit card purchase	$r = .133, p < .0001$
When entering an event like a professional football game	$r = .134, p < .0001$
When boarding a plane if one in five passengers were required to go through additional screening due to a false biometrics reading	$r = .153, p < .0001$

technologies that regulate human behavior, is higher. The prevention of terrorism is one such example. In the post–September 11 environment of heightened risk of terrorism, biometric technology was painted as a potential solution to identify individuals who might be classified as terrorists. This may explain why there is an especially high correlation in the survey data between the threat of another terrorist attack and the proposed use of biometric identifiers to ensure identity on terrorist watch lists, identification of guest workers, and a background check on foreigners. This risk is not easily managed by the particularized knowledge of the individual, justifying for the respondents' paternalistic intervention that made use of biometric identifiers.

Another type of risk that influenced the acceptability of biometric identifiers was identity theft. In its 2005 report on identity theft, for example, the Federal Trade Commission noted these results (http://www.ftc.gov/os/2007/11/SynovateFinalReportIDTheft2006.pdf):

• 1.4 percent of survey participants, representing 3.2 million American adults, reported that the misuse of their information was limited to the misuse of one or more of their existing credit card accounts in 2005. These victims were placed in the "Existing Credit Cards Only" category because they did not report any more serious form of identity theft.

• 1.5 percent of participants, representing 3.3 million American adults, reported discovering in 2005 the misuse of one or more of their existing accounts other than credit cards—for example, checking or savings accounts or telephone accounts—but not experiencing the most serious form of identity theft. These victims were placed in the "Existing Non-Credit Card Accounts" category.

• 0.8 percent of survey participants, representing 1.8 million American adults, reported that in 2005, they had discovered that their personal information had been misused to open new accounts or to engage in types of fraud other than the misuse of existing or new financial accounts in the victim's name. These victims were placed in the "New Accounts & Other Fraud" category, whether or not they also experienced another type of identity theft.

Perhaps because of the number of individuals affected by identity theft and the awareness of the threat, there were positive relationships between respondents' levels of concern about identity theft and their

Table 6.3
Correlations between respondents' reported level of concern about identity theft and their perceived acceptability of using biometrics in various situations

Biometric uses	Correlation coefficients
When purchasing a gun	$r = .132, p < .0001$
To check against a terrorist watch list	$r = .128, p < .0001$
To do a background check on all foreigners	$r = .201, p < .0001$
As part of identification for guest workers	$r = .177, p < .0001$
When withdrawing money from a bank	$r = .134, p < .0001$
When boarding a plane if one in five passengers were required to go through additional screening due to a false biometrics reading	$r = .134, p < .0001$

perceived acceptance of using biometrics in the circumstances of purchasing a gun; checking against a terrorist watch list; background checks on foreigners; identification of guest workers; and boarding a plane even if one in five passengers was required to go through additional screening (table 6.3).[3]

In addition to the issues of identity theft and the threat of another terrorist attack, the control of illegal aliens tended to support the use of biometric identifiers. There was a positive relationship between respondents' perceived level of concern with illegal aliens in the United States and the perceived acceptance of using biometrics to do a background check on all foreigners ($r = 0.250$, $p < 0.0001$) and as part of identification for guest workers ($r = 0.200$, $p < 0.0001$). As respondents' level of concern with illegal aliens increased, so did the reported acceptability of using biometrics. These risks, outside the control of the individual, influenced the social acceptance of biometric identifiers and would indicate societal support for programs such as the US-VISIT program.

These risks, however, did not justify all forms of paternalistic intervention using biometric systems of identification.[4] Paternalistic intervention becomes more difficult to justify when there is interference with the liberty of individuals for the prevention of a harm that is not immediate and direct to them. Also problematic is the concern that the "protection" achieved with paternalistic intervention may benefit one class of individuals while at the same time interfering with the liberty of other individuals whose interests are not implicated. Information technologies target not only the guilty or the wanted. As Dworkin (1971) noted, the idea of paternalism is complicated in its application:

> In "pure" paternalism the class of persons whose freedom is restricted is identical with the class of persons whose benefit is intended to be promoted by such restrictions. . . . In the case of "impure" paternalism in trying to protect the welfare of a class of persons we find that the only way to do so will involve restricting the freedom of other persons besides those who are benefitted. . . . Paternalism then will always involve limitations on the liberty of some individuals in their own interest but it may also extend to interference with the liberty of parties whose interests are not in question. (111)

Consider the difficult case of combating terrorism with the use of personal information to identify terrorists or the use of surveillance to spot dangerous activity and the potential loss of liberty that can result for those who are not terrorists. The prevention of terrorism involves the mitigation of risk for society at large, but it may also impose potential costs on the liberty of those who in no manner pose a danger. Unlike the easy cases Mill recited—saving an individual from falling off a bridge, for example—a liberty interest is implicated when prevention of terrorism is pursued through technologies that regulate human behavior. The use of personal information for the purposes of identifying potential terrorists or general surveillance comes with costs. For many, the loss of public anonymity, or what Slobogin (2006) described as public privacy, is a profound loss of liberty. The psychological and behavioral effects of surveillance can undermine the very essence of liberty. According to Slobogin, surveillance, or even the perception of it, is a powerful force that can undermine the value of public anonymity, resulting in conformity and oppression. It is the perceived constraint on possible choices, in addition to the actual constraints, that represents the potential loss of liberty with the use of personal identifiers or the presence of surveillance technologies. And it is more than the presence of control over information that is problematic. There is also the possibility of erroneous judgment or misidentification for which there may not be a procedural guarantee. The paternalistic act in this instance paradoxically becomes a potential threat to the liberty interests of individuals because, unlike falling off a bridge, the choices that they want to make may be circumscribed in the name of preventing a societal harm. In instances when the liberty interests of individuals and the paternalistic action taken to prevent harm are potentially at odds, or perceived to be at odds, the acceptance of paternalistic intervention treads on delicate philosophical ground.

On the one hand, it does not make sense to assume that if biometric identifiers are used to regulate human behavior under the auspices of paternalistic intervention, there is a concomitant loss of liberty. Increasing reliance on personal information as currency in light of threats to liberty indicates that some form of paternalistic intervention may be necessary from the viewpoint of society. In other words, in a slightly ironic sense, individual liberty may be better preserved with the development of a technological, institutional, and policy framework that can make use of biometric identifiers that regulate human behavior to protect liberty without posing a threat to it. In short, the liberty interests of individuals can no longer be assumed to be inconsistent with the regulatory influence of biometric identification systems.

On the other hand, contending that individuals cannot always act in their best interest with regard to certain risks and that biometric identification is inconsistent with individual liberty does not remove the necessity of determining the proper balance between the two. The enduring question continues to be the appropriateness of paternalistic intervention and its relationship to the preservation of individual liberty. This balance turns on the moral autonomy of the individual and its preservation in light of the moral justification of the paternalism. Clearly, paternalism, in its pursuit of a common good or its avoidance of an evil, should not be unchecked. As H. L. A Hart (1963) argued, "Legal enforcement bears on those who may never offend against the law, but are coerced into obedience by the threat of legal punishment. This rather than physical punishment is what is normally meant in the discussion of political arrangements by restrictions on liberty. . . . The unimpeded exercise by individuals of free choice may be held a value in itself with which it is *prima facie* wrong to interfere" (21). Paternalism—"the protection of people against themselves"—according to Hart, is necessary to the preservation of society, but the enforcement of morality requires justification that society finds acceptable lest the exercise of power be viewed as illegitimate. Hart encouraged "us to consider as values, for the sake of which we should restrict human freedom and inflict the misery of punishment on human beings, things which seem to belong to the prehistory of morality and to be quite hostile to its general spirit" (83). Hart's admonition is important to an understanding that social institutions must be responsive to societal sentiment so that paternalistic power is not perceived as coercive. As he cautioned, "To use coercion to maintain the moral *status quo* at any point in a society's history would be artificially to arrest the process which gives social institutions their value" (75). Striking this balance between the morally permissible paternalistic intervention and the morality of individual autonomy begins with an assessment of societal perceptions.

Morally Permissible Paternalism and Its Limits

There are several dimensions to the evaluative process of permissible paternalistic intervention even in the face of risks that cannot be managed competently by the individual. Paternalistic intervention, because it carries with it the connotation of preserving well-being or preventing some harm, invokes a moral justification that is perceived as sufficient to overcome the liberty interest of the individual or, alternatively, does not contradict it. Moral permissibility is only the first consideration, however. A related consideration is whether there is a duty to intervene to prevent harm. Not unlike Mill's admonition to protect individuals against the harm of selling themselves into slavery, there are instances when paternalistic intervention is not only morally permissible but a duty of the government. As VanDeVeer (1986) asserted:

> Our central question in this inquiry is whether certain types of paternalistic acts or particular paternalistic acts are morally permissible. Those claiming that a paternalistic act is wrong deny its permissibility. If it can be established that an act is permissible, there is always the further question of whether it is impermissible not to do it. If and only if it is true that both an act is permissible and it is impermissible not to do it is there a duty to do it. So, "the" question of justification of paternalism can be understood as two questions: (1) whether paternalism is permissible or (2) whether paternalism is a duty. (40)

The control of pollution, the requirement of vaccinations, the mandate of food and drug testing, and fighting terrorism represent paternalistic interventions that society has come to accept as morally permissible and, more important, for understanding their regulatory effects as even a duty of the government. While some of these examples might represent easy cases on the surface, the societal acceptance of paternalistic intervention is not given wholesale nor does it manifest overnight. Societal changes, which give rise to emerging policy areas ripe for paternalistic intervention, including the increasing reliance on personal information in the current societal context, are evolutionary rather than revolutionary manifestations that require policy and technological innovations.

The normative acceptance of paternalistic intervention is only the first step. Individuals do not give up liberty without a fail-safe. Given the wide range of individual choice and liberty that must be maintained in contemplation of paternalistic intervention, moral justifications must be tempered by limitations. Although a legal right to individual autonomy in relationship to personal information may be ill defined in many circumstances, there still remains moral justification of individual liberty that should serve to limit the reach of morally permissible paternalistic intervention. Samuelson (2000) described the idea of a moral right in personal information as powerful:

> A moral right-like approach might be worth considering as to personal data. As with the moral right of authors, the granting of a moral right to individuals in their personal data might protect personality-based interests that individuals have in their own data. The mixture of personal and economic interests could be reflected in the right. The integrity and divulgation interests may be the closest analogous moral rights that might be adaptable to protect personal data. An individual has an integrity interest in the accuracy and other qualitative aspects of personal data, even when the data are in the hands of third parties. An individual also has an interest in deciding what information to divulge, to whom and under what circumstances.

The moral weight given individual autonomy and the assessment of its importance as a point of limitation to the use of biometric identification systems was evident in the national survey and focus groups. Individuals were not willing to submit to the use of biometric identification systems without some level of concern and desire for limitations. For example, there was a positive correlation between the level of concern that private industry might use their biometrics to track their activities and their level of concern with various threats to data security, including revelation of personal information to the government without permission, revelation of personal information to private companies without permission, and someone getting into their bank account without permission. While the acceptance of biometric identifiers is influenced by the threats of identity

Table 6.4
Correlations between respondents' reported level of concern that private industry might use their biometrics to track activities and their level of concern with various threats to data security

Issues	Correlation coefficients
Someone revealing personal information to the government without permission	$r = .185, p < .0001$
Someone revealing personal information to private companies without permission	$r = .213, p < .0001$
Someone getting into their bank account without permission	$r = .149, p < .0001$

Table 6.5
Correlations between respondents' reported level of concern that their biometric information may be stolen and their level of concern with various threats to data security

Issues	Correlation coefficients
Another terrorist attack within the United States	$r = .201, p < .0001$
Identity theft	$r = .190, p < .0001$
Someone revealing personal information to the government without permission	$r = .259, p < .0001$
Someone revealing personal information to private companies without permission	$r = .247, p < .0001$
Someone getting into their bank account without permission	$r = .225, p < .0001$
Loss of civil liberties as part of the war on terror	$r = .281, p < .0001$
Illegal aliens in the United States	$r = .171, p < .0001$

theft, the potential of another terrorist attack, and control of illegal aliens, the threat of misuse of biometric identifiers continued to operate as a point of concern and an obstacle to societal acceptance (table 6.4).

There were also positive relationships between respondents' reported level of concern that their biometric information may be stolen and their level of concern with various issues. As respondents' reported level of concern with stolen biometrics increased, so did their level of concern with various threats to their data security. For individuals who feared that their biometric information may be compromised, the potential for misuse, ranging from revealing information without permission to a loss of civil liberties, rose. In these examples, the lack of protection afforded to biometric identifiers in terms of regulation tended to undermine the protection these identifiers offered against terrorist attacks and the threat of identity theft. Moreover, the concern that biometric information might be stolen correlated with a fear of a loss of civil liberties and a misuse of personal information. Table 6.5 outlines the statistically significant relationships.

The Limits of Paternalism

The acceptance of morally permissible paternalistic intervention using biometric identification systems is undermined by fears of misuse and a lack of protection for the loss or theft of biometric identifiers. There is a clear societal imperative that biometric technology, which can serve the purpose of managing risks, must be subject to safeguards to ensure good judgment and limit the regulatory effect on individual liberty.

The management of risks involves making a judgment on behalf of another individual, which may or may not be in line with the individual's operative preference, intention, or disposition but which is done for the purpose of promoting a benefit or preventing harm because the individual either does not have the capacity to exercise the correct choice or has insufficient knowledge to make the choice. Tasking government with the management of risks, which makes use of biometric information, introduces questions regarding the methods of ensuring good judgment. "Governments often function as third parties which, with protective aims, intervene to delimit the risks faced by citizens who are subjected to risks as a result of others (other individuals, organizations, or Mother Nature)" (Vandeveer 1968, 318).

The duty assumed by the government to prevent harm by making use of information, whether it is in the form of biometric identification systems or other forms, in pursuit of the prevention of identity theft or fighting terrorism is not unlike the duty of the doctor to protect the patient from harm by controlling information. In each case, a judgment is made on behalf of the individual to protect against risks that he or she is factually incompetent to handle. There is also a similarity in the duty that is owed and the responsibility that is given to both the physician and government to protect the interest of the patient or, in the latter case, society by the use and control of information. Buchanan (1983) described the judgment exercised as difficult: "We begin to see the tremendous weight that this paternalistic argument places on the physician's power of judgment. She must not only determine that giving the information will do harm or even that it will do great harm. She must also make a complex comparative judgment: Will withholding the information result in less harm on the balance than divulging it?" (66).

The dilemma of judgment is difficult for both the physician and the government. The use of information—either withholding it or divulging it—is undertaken with the goal of preventing harm or achieving a societal good. Yet in both cases, a harm might occur because of the preventative action taken. Using the example of a patient with terminal cancer, Buchanan (1983) explained, "The physician must not only determine that informing the patient would do great harm but that the harm would be greater on balance than what harm might result from withholding information" (66). The withholding or control of information in the case of a physician-patient relationship is based on the premise that the physician is under an obligation to do no harm to a patient. The calculation involves the withholding of information, justified in terms of preventing harm: "Paternalism is the interference with a person's freedom of action or freedom of information, or the dissemination of misinformation, or the overriding of a person's decision not to be given information, when this is allegedly done for the good of that person" (Buchanan, 1983, 62). Whatever the evidence for the judgment, the doctor is given a duty to assess what will be best for the patient. The patient has to rely on the judgment of the physician and may or may not know if he or she is privy to all information relevant to the decision. Guidance for this judgment includes these ideas:

The physician's duty—to which she is bound by the Oath of Hippocrates—is to prevent or at least to minimize harm to her patient.

Giving the patient information X will do great harm to him.

Therefore it is permissible for the physician to withhold information X from the patient.

In the case of biometric identifiers, paternalistic intervention to prevent terrorism, address identity theft, or combat fraud represents control over personal information that is subject to misuse or lack of good judgment by those institutions that possess it. This act of judgment as a means of ensuring the respect for the moral autonomy of the individual is akin to that exercised by medical doctors in the care of their patients and similarly requires guidance.

The question is how to ensure the good judgment of institutions. The answers from the survey respondents are revealing. A consideration of the types of limits that would be in place to counteract erroneous judgments and ensure the protection of personal information was important to the management of judgment as it pertained to the use and prevention of misuse of biometric identifiers. When asked about their comfort levels with biometric identifiers in the context of policies to protect them, participants responded positively (table 6.6).

Table 6.6

Reported comfort levels with various biometric policies (N = 1,000)

Policies and Procedures	A lot more comfortable (%)	Somewhat more comfortable (%)	No more comfortable (%)
There was a written privacy policy specific to biometrics (e.g., a privacy policy from a credit card company).	18.6	49.3	29.8
Biometrics were used only for a purpose that was clearly explained.	29.9	49.7	19.0
Respondents' information was shared only with people they were aware of (such as when applying for a credit card).	29.9	43.3	25.8
Companies were required to get respondents' permission before selling, leasing, or disclosing personal information.	45.2	32.5	21.3
There were procedures in place to allow the respondents to check the accuracy of the information attached to their biometric identifier.	41.0	41.3	16.3
Biometrics were used in accordance with existing privacy guidelines such as those laid out by the U.S. Constitution.	30.7	46.2	21.0

The majority of respondents were significantly more comfortable with the use of biometrics if permission was given before selling, leasing, or disclosing personal information. Consent as a means of ensuring autonomy was an important limitation on the judgment exercised over the control of biometric information. But consent was not the only factor in ensuring the protection and respect for moral autonomy, especially when respondents were willing to cede discretion to an institution to make a judgment on their behalf. Procedural guarantees to ensure accuracy of information, accordance with other legal protections, and limitations on the sharing of biometric identifiers were also considered significant. Specific laws that serve to protect biometric information were also essential to the respondents (figure 6.1). For example, 87 percent of respondents favor state and federal privacy laws that would apply specifically to protecting biometric data. There was a statistically significant difference between respondents who favored state and federal privacy laws that apply specifically to protecting biometric data and those who did not favor these laws with respect to their perceived level of importance of the privacy of their personal information ($t(970) = -2.50$, $p < 0.0001$). Respondents who favored specific state and federal biometrics privacy laws reported a higher level of importance of the privacy of their personal information (mean =

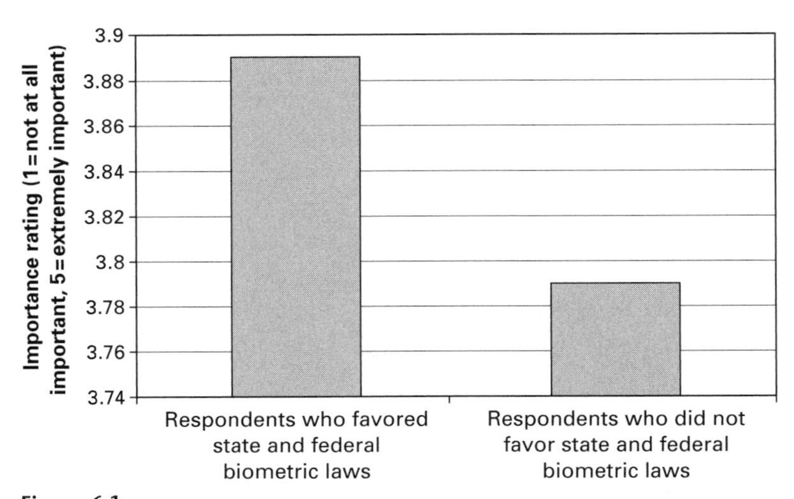

Figure 6.1
Average reported level of importance of the privacy of personal information

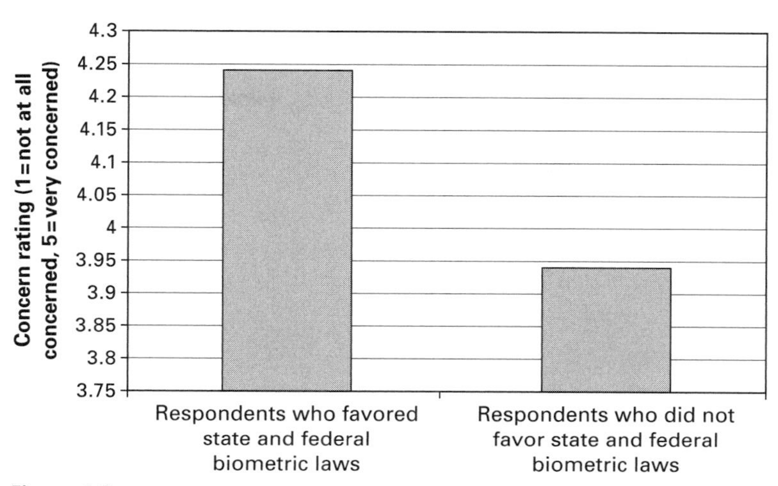

Figure 6.2
Average reported level of concern with another terrorist attack

3.89, sd = 0.37) than respondents who did not favor such laws (mean = 3.79, sd = 0.52).

The threat of another terrorist attack did not eliminate the perceived need for protections for biometric information (figure 6.2). There was a statistically significant difference between respondents who favored state and federal privacy laws that apply specifically to protecting biometric data and those who did not favor these laws in regard to their level of concern with another terrorist attack within the United States ($t(914) = -2.46$, $p < 0.0001$). Respondents who favored specific state and federal biometric privacy laws reported a higher level of concern with another terrorist attack on the United States (mean = 4.24, sd = 1.14) than respondents who did not favor such laws (mean = 3.94, sd = 1.45). Even if individuals were concerned about another terrorist attack and favored the use of biometric identification systems, they also favored protections for the personal information.

Interestingly, there was also a statistically significant difference between respondents who favored state and federal privacy laws that apply specifically to protecting biometric data and those who did not favor these laws in regard to their level of concern with illegal aliens in the United Sates ($t(964) = -2.10$, $p < 0.0001$) (figure 6.3). Respondents who favored specific state and federal biometric privacy laws reported a higher level of concern

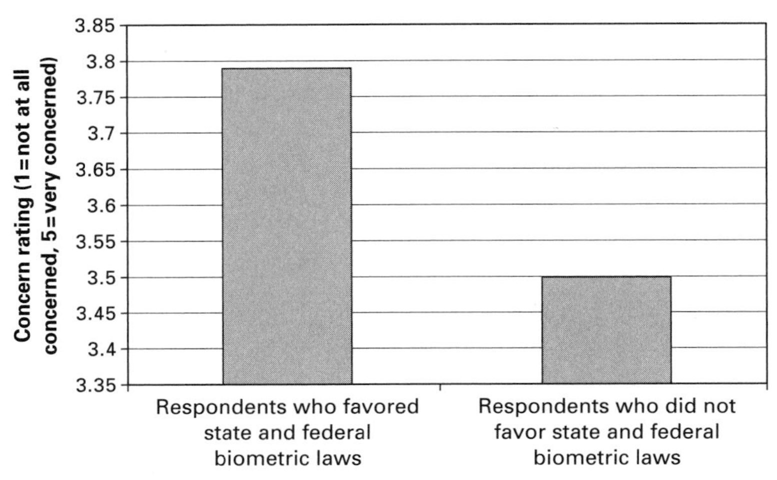

Figure 6.3
Average reported level of concern with aliens

with illegal aliens in the United States (mean = 3.79, sd = 1.33) than respondents who did not favor such laws (mean = 3.50, sd = 1.57).

The interest in protecting personal information extended to other forms of identification as well. There were positive relationships between respondents' perceived protection of their personal information by state and federal laws and how safe they reportedly felt using various identifiers to ensure that only they have access to personal records. As respondents' perceived level of protection increased, so did their perceptions of the level of safety offered by various identifiers. Table 6.7 shows the statistically significant relationships.

September 11: The Call of the Sirens?

Are individuals in society like Odysseus? Have we surrendered our individual liberty because of the risks that we face? Recall that as Odysseus and his crew sailed past the Isle of the Sirens, Odysseus ordered his crew to bind him to the mast so that he would not be tempted by the seductive songs of the Sirens and jump overboard to be inevitably killed: (Homer 1963, 214).

Some would describe the influences of September 11 and the information age on the perception of risk and greater support for biometric

Table 6.7
Correlations between respondents' perceived protection of their personal information and perceived safety level using various identifiers for the protection of their information

Identifiers	Correlation coefficients
Social security number	$r = .228$, $p < .0001$
Driver's license number	$r = .194$, $p < .0001$
Mother's maiden name	$r = .141$, $p < .0001$
Date of birth	$r = .183$, $p < .0001$
Street address	$r = .152$, $p < .0001$
City	$r = .126$, $p = .0001$
Credit card number	$r = .140$, $p < .0001$
Bank account number	$r = .153$, $p < .0001$
Phone number	$r = .172$, $p < .0001$
Occupation	$r = .136$, $p = .0001$
Home telephone number recognized by an automated security system	$r = .158$, $p < .0001$

identifiers as a powerful ideological force akin to the call of the Sirens. Jeffrey Rosen (2005), for example, argued that a crowd psychology reigns in a world affected by the events of September 11 and already altered by the preponderance of information technologies: "People believe that they are most likely to be victimized by the threats of which they are most afraid. . . . Terrorism, in this regard, is among the risks that are so horrific to contemplate and hard to control that people's judgments about its probability are likely to be especially clouded by their fear of the outcome" (74).

Rosen asserted that there is an irrational and emotive acceptance of the use of invasive surveillance technologies in order to obtain a sense of security even at the expense of a loss of liberty. In addition, he argued, because of a crowd mentality, individuals do not in fact make a cost-benefit analysis in the way they should. Yet do individuals really accept the presence of surveillance without consideration of limits? Do we trade places with Odysseus, tied to the mast of our ship, because we cannot control ourselves in light of the risk of the Sirens? Individuals vest power in democratic governance and yield to it a certain degree of autonomy, but it does not follow that the events of September 11 and the increasing presence of information technologies in our lives have converged to lead to an irrational and emotive quality in the American public or, worse yet, a fearful crowd. In fact, respondents in both the focus group and survey suggested that paternalistic intervention using biometric identification systems is not without its limits.

Although there is no expectation that every form of paternalism involving use of biometric identifiers as protection against threats must be tempered, neither is there a rush to be tied to the mast. There is, in other words, an important place for limitations. Individuals are willing, in certain circumstances, to acquiesce to the use of biometric technology, but not without a consideration of how to ensure continued respect for their individual autonomy. While the moral imperative is important to the acceptance of paternalism, it is also essential that limitations be in place to check those potential "immoralities" of paternalism to preserve the values, norms, and expectations that it triggers. This finding likely extends beyond biometric technologies and represents only one step in the methodological practice of evaluating the evolving nature of technology and society. Technological innovation influences human behavior, and human behavior influences technological innovation.[5] Technology is never value free and can be both regulative and constitutive of human behavior. The objective is to determine how and when technologies that regulate can serve society and not undermine society's values, norms, and expectations.

In evaluations of the structural and substantive consequences of information technologies, including biometric technologies, it is not enough to ask whether the action taken reaches the desired regulatory effect or whether the use of the technology is efficient or reliable. The greater question is whether the regulating effects of technology respect the fundamental values, norms, and expectations of society. The question is not only about preventing the regulatory effect of technologies, but, in certain circumstances, using the regulatory effect of technologies as a tool to protect the moral community: "In a community of rights, the fundamental question is whether the technology threatens to change the cultural environment in a way that no aspirant moral community can live with. If there is a real concern that the technology presents such a threat, regulators, as stewards of the moral community, should no further with that kind of strategy" (Brownsword 2008, 25).

The challenge is how to create a methodology for incorporating the relevant norms, values, and expectations of society into the development and deployment of technology. While functionality and efficiency are values that usually are front and center in the development of technology, the incorporation of values of a more humanistic dimensions is more difficult for several reasons. (Flanagan, Howe, and Nissenbaum 2008) Most obvious, the sources of values, norms, and expectations are ambiguous given the wide variety of philosophical, traditional, legal, and other sources. As an alternative, empirical investigations, like the one done here, provide a source for understanding the norms, values, and expectations that are important for the design and implementation of technology, broadening the findings with ideological context developed from philosophy, tradition, and law. Ideas of individual liberty are not formed *tabula rasa*; they reflect a socialized understanding that can also indicate points of change and evolution.

7 Conclusion

Societal acceptance of biometric technology involves conflicting considerations. Individuals may accept biometric systems of identification for the purpose of identity assurance in an array of circumstances and in a variety of public and private institutional settings, while having concerns about the potential loss of liberty that might occur with the misuse of biometric identifiers. Biometric systems of identification can serve the purpose of protecting the currency of personal information with identity assurance, but the development and deployment of biometric technology must also acknowledge the limits of individual liberty.

The implications of biometric systems of identification for individual liberty requires a reconsideration of liberty that acknowledges an exchange of the currency of personal information, for which biometric data as a form of identity authentication serves an important role. Individual liberty must be reconceptualized to account for the use of data by individuals for communication, transactions, and networking. Data minimization has long been advocated as a means by which to protect individual liberty, but it may, in fact, impede it. While individuals rely on the exchange of personal information as expressive of their liberty, they also face the risks of identity theft, surveillance, and fraud. As the survey and focus group data explored in the previous chapters show, protecting the currency of personal information while preserving the norms, values, and expectations of society can be accomplished.

Achieving a satisfactory balance requires rethinking the terms. For example, to call the individual liberty interest in personal information only a "privacy" interest implies that disclosure should be minimized. One of the clear changes in our norms, values, and expectations is that we use personal information as a currency for our interactions in the public and private sectors. Rather than limiting the liberty interest to a definition of privacy, the trend toward data empowerment requires the goal of preserving decisional autonomy in the revelation of personal information instead of the requirement of data minimization. Decisional autonomy requires that, whenever possible, the individual be allowed discretion to divulge information or to remain anonymous. Decisional autonomy is threatened not only by the misuse of personal information but also by surveillance that tracks human behavior with regulatory information technology, by either physical observation or the covert attendance on transactions or communications. Decisional autonomy does not exist in a vacuum; in certain circumstances, it must be limited either because the individual lacks the requisite knowledge to protect his or her interest or, alternatively, the individual interest might run counter to the common good. In the circumstances that threaten the currency of personal information, such as identity theft or fraud, decisional autonomy may be limited by paternalistic intervention in the form of policy or the regulatory presence of information technologies to combat harmful behavior, that is, to reduce fraud or to ferret out identity theft. Yet regulatory technologies must strive not to threaten individual liberty as in the case of unnecessary surveillance of communications, transactions, or physical activity. This balancing requires that the enduring principles of liberty coexist with regulation or technological regulatory tools. Setting aside data minimization as the only means of protecting the liberty interest in personal information means institutional protections to promote data empowerment.

Increasingly public and private sector institutions are serving as the foundation of a civil society and associational life that is built on the exchange of information. These institutions must employ technological and policy safeguards to protect personal information. These safeguards might include policy mandates, which demand consumer choice, transparency, and notices of disclosure or procedural protections. Technological innovations that protect information acquired by the public and private sectors are also important and may include an array of technological tools, ranging from identity assurance to surveillance. These regulatory efforts, whether regulating by technology or policy, do not require the demise of individual liberty. To the contrary, individual liberty depends on protecting the exchange of personal information. Individuals need not be victims in the face of information technologies; they can be and are participants. Arguments about preserving liberty must account for a changing conception of the relationship that society has with information technology and the information it generates. Acknowledging this change, however, does not mean that traditional norms, values, and expectations cease to matter to individuals. Individuals will neither give up their reliance on personal information as currency nor will they forget about concerns of protecting their liberty.

Biometric technology is at the center of this evolution. Biometric identification systems stand to increase the safety of interactions that use personal information by reducing fraud, identifying potential terrorists, and enhancing the efficiency of interactions with identity assurance. At the same time, the biometric data generated must not threaten the norms, values, and expectations of individual liberty with the prospects of unnecessary surveillance of activities, communications, or transactions. Although some might argue that societal norms, values, and expectations lead us down the slippery slope of the reformulated concerns of the tyranny of the majority, especially in light of September 11 and the information age, this research has revealed an American public that is far more thoughtful.

Insights

It is important to reiterate that the American public has not set aside the traditional values of individual liberty because of the influences of September 11 or the information age. Yet traditional values of individual liberty have been transformed by the growing reliance on personal information and are influenced by the threats that make use of it and undermine its value. For example, individuals balk at generalized surveillance at public events or identity assurance that takes the place of personal relationships, as in the case of picking up a child at day care. At the same time, concerns of identity theft and terrorism increase the societal acceptance of paternalistic intervention to regulate human behavior with technological tools, such as biometric systems of identification. Requiring fingerprints of individuals coming into the United States, as in the US-VISIT program, is one regulatory function that has garnered societal acceptance. Although there are circumstances when society is more willing to accept identity assurance in the public and private sectors, it is not without some continued approbation. Individuals continue to worry that their biometric data might be used for surveillance of unnecessary activities or put to unintended purposes. Societal acceptance of biometric systems of identification varies according to institutional uses and policy purposes, and, as noted, confidence in public and private sector institutions varies across the same spectrum.

In the study examined in this book, different levels of confidence tended to hinge on either the type of procedural safeguards that were in place or the policy objectives that justified the use of biometric systems of identification. For example, the objectives of fighting terrorism by the Department of Homeland Security or combating tax evasion by the Internal Revenue Service generated more confidence because of the institutional objectives for personal information. Beyond a well-defined administrative purpose, individuals were more confident in institutions that had legal procedural protections in place to protect personal information. The highest ratings of confidence were given to medical providers and financial institutions in both the focus groups and the national survey. The focus group participants said that legal protections that exist in the Health Insurance Portability and Accountability Act (HIPAA) and Gramm-Leach-Bliley Act (GLBA) made a difference in their confidence in medical and financial institutions.[1] This is related to the clear finding in the focus groups and national survey that procedural protections for the deployment and development of biometric systems of identification would increase the comfort level of individuals with biometric technology. These procedural protections included a limitation on use and sharing of information; required permission before selling, leasing, or disclosing biometric information; and allowed respondents to check the accuracy of the information attached to their biometric identifier.

Perhaps the finding that legal protections created confidence in institutions that handled personal information is not so surprising. We are socialized to believe that law protects our rights and liberties: this is, after all, the Enlightenment promise of a rational system of law. The primary source of our normative beliefs about privacy, anonymity, decisional autonomy, and limits on paternalistic intervention is informed by the legal principles that have given them life in our political past. The symbolic effect of law and policy is not to be taken lightly. Individual liberty, which is protected with legal guarantees as in the case of HIPAA and Gramm-Leach-Bliley, actually results in greater compliance by institutions and their use of personal information. The mandates of disclosures, notice requirements, limitation-of-use principles, consent provisions, and restrictions on the sharing of information give substance to individual liberty and the normative values of privacy, decisional autonomy, and limitations on paternalistic intervention. These protections serve to build trust and confidence in the medical and financial institutions, which make use of the personal information providing benefits for legitimacy of institutions. As Tyler (2006) stated,

> Legitimacy is also an issue on the group, organizational, or system level, where the legitimacy of authorities and institutions is a part of the overall climate or culture. Discussions of the stability of social and political systems have long emphasized the importance to effective governance of having widespread consent from those within the system. Such widespread consent enables the more effective exercise of social and political authority, since authorities can appeal to the members based upon their shared sense of values. (2)

There are many examples from which to construct applicable policy safeguards for institutional use of biometric data. For example, the first code of fair information practices, developed in 1973 by an advisory committee in the Department of Health, Education and Welfare provided a core statement of principles. These principles set out basic rules designed to minimize the collection of information, ensure due-process-like protections where personal information is relied on, protect against secret data collection, provide security, and ensure accountability. In general, the principles emphasized individual knowledge, consent, and correction, as well as the responsibility of organizations to publicize the existence of a record system, ensure the reliability of data, and prevent misuse of data. This code formed the basis for future fair information practices in the public and private sectors, as well as future U.S. laws (including the Privacy Act passed by Congress in the following year). The basic principles of the code reflected modern privacy concerns: no personal data record-keeping systems could exist in secret; individuals must be able to discover what information is in their files, and to what use it is being put; individuals must be able to correct inaccurate information; organizations creating or maintaining databases of personal information must ensure the reliability of the data for its intended use and prevent misuse; and individuals must have the ability to prevent the use of information for purposes other than that for which it was collected.

These principles were also reflected in the Organization for Economic Cooperation and Development (OECD) Guidelines of Privacy and Transborder Flows of Personal Data in 1980.[2] The OECD guidelines create standards that served as the principles for privacy policies and are the foundation of current international agreements and self-regulation policies of international organizations. The content of the guidelines is somewhat indefinite but provides an indication of a framework for protecting privacy in light of biometric technologies. The OECD code emphasized eight principles: collection limitation, data quality, purpose specification, use limitation, security safeguards, openness, individual participation, and accountability. Under OECD guidelines, the collection of personal data is limited; data are obtained by lawful means and with consent where appropriate. Data must be relevant to the purpose intended and must be accurate and up to date. The purposes of collection must be specified at the time of collection or earlier, and use must be limited to those explicit purposes or other related purposes. Consent of the subject or by the authority of the law also limits disclosure. Additionally, security safeguards are required against risk of loss or unauthorized access, modification, or disclosure. A general policy of openness, designed to establish the existence and nature of personal data, their purposes, and the identity of the data controller, is in place as well. There are further requirements that individuals should be able to obtain confirmation of the existence of their personal data records, have access to that information or be given adequate reasons if access is denied and be able to challenge the denial of access, and to challenge the existence or correctness of data. Finally, data controllers are expected to be accountable for enforcing these principles.

The Federal Trade Commission has relied on these principles to argue that notice/awareness, choice/consent, access/participation, integrity/security, and enforcement/redress are central to the development of privacy policy.[3] These principles also inform the privacy principles articulated at the conferences of the Asia-Pacific Economic Cooperation (APEC).[4] At the 2003 Singapore conference, twenty-one member states began to formulate a set of privacy principles in an attempt to achieve a balance between informational privacy and government interests. Reflective of a utilitarian framework approach to privacy, the APEC draft aimed to prevent harm while maximizing benefit in society. With these two counterpoints in mind, privacy protections are to be designed to realize the benefits of both information flow and individual control.

The APEC principles mirror those of the OECD. Security of information, safeguards against access or disclosure, and proportionality of harm and sensitivity of the information are considered. Notice of collection, purpose, and disclosure, as well as the identity and contact information of the data controller, are standard operating procedures, while the principle of choice gives data subjects the right to know what options are available to them in limiting the use or disclosure of their information.[5] Other limits on the actions of data controllers in the APEC framework include those on collection and use: collection must be relevant to the purpose and use limited to the purpose of collection unless the consent of the data subject is given, a service is being provided to the subject, or the use is under authority of law. Additionally, accuracy of information is required, as is the ability of data subjects to obtain, correct, complete, or delete stored information about them.

The European Directive on Data Privacy represents another approach.[6] The first principle of the directive requires that information be collected only for specific and specified purposes, used only in ways that are compatible with those purposes,[7] and stored no longer than is necessary for those purposes.[8] As a secondary check on the misuse of information, the directive requires that measures that are appropriate to the risks involved be taken to protect against the "alteration or unauthorized disclosure . . . and all other unlawful forms of processing."[9] Particular types of data receive special protection under the directive. According to its language, "racial or ethnic origin, political opinions, religious or philosophical beliefs, . . . and . . . data concerning health or sex life" are generally forbidden.[10] With regard to individual control, the directive requires that processing activities "be structured in a manner that will be open and understandable."[11] Similarly, personal information cannot be transferred to third parties without the permission of the subject.[12] The directive also requires independent oversight and individual redress.[13] These principles ensure a right to access personal information, correct inaccurate information, and pursue legally enforceable rights against data collectors and processors that fail to adhere to the law. Individuals have recourse to the courts or government agencies to investigate and prosecute noncompliance by data processors.

These principles offer guidance on the potential frameworks from which to choose in order to construct policy safeguards for protecting personal information like biometric that must be secure in order to provide identity assurance for other personal information. But for some, fundamental omissions undermine the value of the principles. For instance, Rule (2007) argued that the ethical justification for information gathering is missing:

> What they do not do is address the central ethical issue implicated in the extension of surveillance: the tension between an essentially utilitarian logic of efficiency and a Kantian logic of rights. There can be no doubt that widening surveillance is efficient for all sorts of institutional purposes—that it helps allocate credit, collect taxes more productively, track would-be terrorists and other wrongdoers, crack down on unlicensed or uninsured drivers, direct advertising to just the most susceptible consumers, and on and on. Were it not for these attractive efficiencies, government and private organizations would never bother to invest the vast sums needed to create systems. But whether the growth of these systems is compatible with values of individual autonomy and choice over "one's own" information is another matter entirely. (27)

As Rule's criticism demonstrates, although guidelines and principles designed to protect and enhance the exchange of personal information are in place, there is a lack of clarity as to whether the benefits reaped by the public and private sector institutions are compatible with the values, norms, and expectations of society. It is similarly problematic to assume that they are not. The values, norms, and expectations associated with personal information are complex and subtle and, more important, are not always at odds with its use in the public and private sectors. It is wrong to assume that if the individual relinquishes his or her information, there is a benefit to be gained by the institution and a requisite loss of choice for an individual. The protection of individual liberty in a civil society that relies on the currency of personal information does not simply translate into isolation or data minimization; instead it requires rethinking how to protect and enhance data empowerment because this is an important part of individual choice over personal information. It is also problematic to assume that the institutional goals of combating fraud, fighting terrorism, or countering identity theft are incompatible with the liberty of the individual.

Individual liberty can accommodate, and even be promoted by, paternalistic intervention, in the form of policy or technology, without a concomitant loss of liberty. Our liberal political tradition relies on the principle of self-determination, which is ensured by allowing the individual to choose the best course of action based on a personal assessment of his or her particularized knowledge and interest. An assumption is that interference with this type of autonomy equates with a loss of liberty. This assumption has influenced the management of personal information, which has long relied on individual consent as a primary means of ensuring the protection of liberty. Certainly consent serves an important role in preserving the liberty associated with individual autonomy, but there are some obvious problems when it comes to personal information as currency. Destabilization of the currency of personal information in the form of identity theft, terrorism, or identity theft must be countered with policy and technological safeguards if the currency of personal information is to continue to hold its value and individual liberty is to be protected. There are circumstances in which an individual is rendered factually incompetent to manage risks and knows it, preferring paternalistic intervention to protect against the risks that he or she is unable to manage. The threats of identity theft or the risk of another terrorist attack are two such circumstances that garnered societal acceptance of biometric systems of identification. In these situations, the individual does not have sufficient knowledge to protect his or her interest and accepts identity authentication as a solution. Clearly paternalistic intervention in the form of regulating technologies, such as biometric systems of identification, cannot be characterized as a loss of liberty. In fact, noninterference in the case of factual incompetence represents a loss of individual liberty because deception or insufficient knowledge can infringe on the ability of the individual to protect his or her interest.

A reconsideration of Odysseus and the Sirens is in order. When Odysseus asked to be tied to the mast of the ship so that he would not be lured to the rocks by the Sirens, he told his men to refuse all subsequent orders to set him free. Once tied, Odysseus struggled against the bonds, and in that moment he was not free to act on the call of the Sirens. Odysseus asked to have his liberty limited to protect himself from the Sirens, a danger or risk that he knew he was incapable of overcoming on his own. The order to his men was a limitation of liberty, but it served to limit the risk that would ultimately undermine his liberty. The request Odysseus made is not unlike paternalistic intervention in the form of regulation to protect against risks that individuals are factually incompetent to manage.

Like Odysseus, individuals may accept limitations on their liberty when they cannot otherwise protect against risks.

Yet the acceptance of paternalistic intervention does not negate the need to hold individual liberty as the end goal. In other words, Odysseus was tied to the mast to protect his liberty against his exercise of misguided choice under the influence of the Sirens. He was restrained in consideration of his future liberty. In this way, individual liberty should continue to operate as a moral limitation on the paternalistic intervention of public and private sector institution. The question is how to conceptualize public policy or technological innovation that can effectively enable the moral goal of personal liberty to constrain paternalism but allow paternalistic intervention when protection is warranted. The answer cannot simply be derived from existing protections, but must be informed by an assessment of the norms, values, and expectations of society. Society would not accept information acquisition or the presence of information technologies if respect for individual liberty did not continue to reign. In fact, protections that mirror societal perceptions of individual liberty will not only protect the value of the currency of personal information but will confer legitimacy on the institutions and the tools of information technologies they use, including biometric systems of identification.

Biometric systems of identification and the institutions that use them are in a position to create new users, but it is not only reliability, accuracy, or efficiency that will win the American public over. This book began with a discussion of the users in the development and deployment of technology and the influences they have had on technology. Users have long transformed technological innovation to suit their purposes, whether by "scratching" turntables or using the telephone for long conversations rather than short ones as the inventors had expected (Fischer 1992). But as Trevor Pinch (2005) has inquired into the origin of users, focusing on the importance of marketers and salespeople in the development of technology, specifically the minimoog—a synthesizer distinguished by its small size, and with a portable keyboard. Pinch wrote that "when it was first developed there was no clear conception within the Moog company who would buy or use this new instrument" (252). However, the efforts of one synthesizer salesman, Dave Van Koevering, transformed the future of the minimoog into an overwhelming success. Biometric systems of identification and the institutions that use them might not necessarily use marketing or salesmanship to secure societal acceptance; an understanding of the political ideology of the potential users might be more important.

Biometric systems of identification in public and private institutions must create users by reflecting the values, norms, and expectations important to them in the form of policy protections or technological safeguards; this includes a consideration of the societal perceptions of privacy, decisional autonomy, anonymity, trust and confidence in institutions, and limits on paternalistic intervention. This is especially true in circumstances of factual incompetence of the individual, such as the prevention of terrorism or the reduction of identity theft, where the institutional objective must be tempered by procedures that continue to respect individual liberty.

Conceptually the requirements of informed consent provide a moral framework for preserving individual liberty. The idea of informed consent has its roots in moral philosophy, the health profession, law, and the social and behavioral sciences. Central to the idea of informed consent is the imperative to preserve "privacy, voluntariness, self-mastery, choosing freely, the freedom to choose, choosing one's own moral positions, and accepting responsibility for one's choices." (Faden and Beauchamp 1986, 7). Different from simple consent, informed consent prompts a discussion about the responsibilities and duties of those individuals and institutions tasked with preserving individual liberty:

> More important is the question of the exact demands the principle makes in the consent context—for example, as to requirements that certain kinds of information be disclosed. Another question concerns the restrictions society might rightfully place on choices by patients or subjects when these choices conflict with other values. If choices might endanger the public health, potentially harm a fetus, or involve a scarce resource for which a patient cannot pay, it may be justifiable to restrict exercises of autonomy severely, perhaps by state intervention. If restrictions are in order, the justification will rest on some competing moral principle such as beneficence or justice. (Faden and Beauchamp 1986, 9)

Informed consent generally requires a process by which a person or institution guarantees disclosure, comprehension, voluntarism, and competence to the individual. These general concepts must be translated into policy directives and directives for the development and deployment of biometric technology. For example, the requirement of disclosure is intended to ensure awareness and assent on the part of the individual. Disclosure that is designed to achieve awareness and assent requires procedures of communication that provide information for the exercise of informed consent. In the patient-doctor relationship, communication relies on the exchange of information dialogically with questions and answers sufficient to build informed consent. In the institutional context, informed consent requires a consideration of the rules and procedures necessary to achieve it. The practice of informed consent as it applies to the control of personal information translates into procedural channels to exert control or to counteract potential misuse of information.

Clearly there is a place for policy and technological protections for personal information, but there is also a need to continue an ethical debate about the justifications and implications of information technologies in relationship to the existing and potential users. The norms, values, and expectations of users must inform the development and deployment of biometric systems of identification.

Appendix A: Safety of Identifiers: Factors of Education

- Date of birth ($F_{(df = 5)} = 5.47$, $p < .0001$)

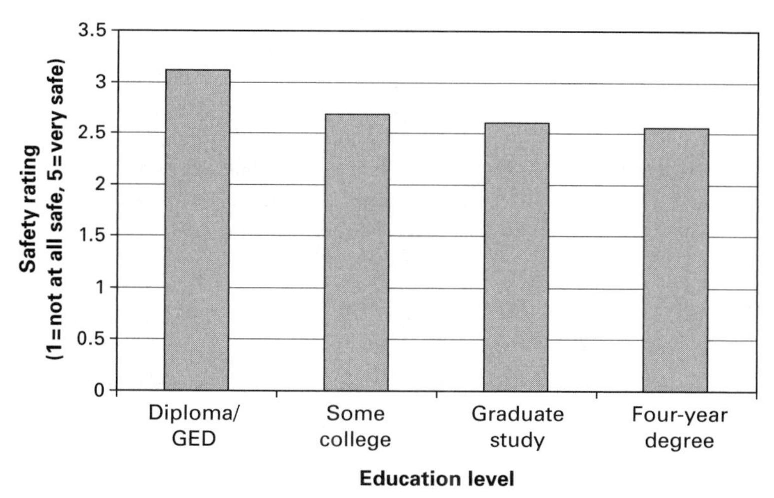

Figure 8.1
Average reported levels of safety with the identifier date of birth. *Note:* Diploma/ GED: mean = 3.11, sd = 1.54; some college: mean = 2.69, sd = 1.42; four-year degree: mean = 2.56, sd = 1.25; more than a four-year degree: mean = 2.61, sd = 1.38.

- Street address ($F(df = 5) = 5.60$, $p < .0001$)

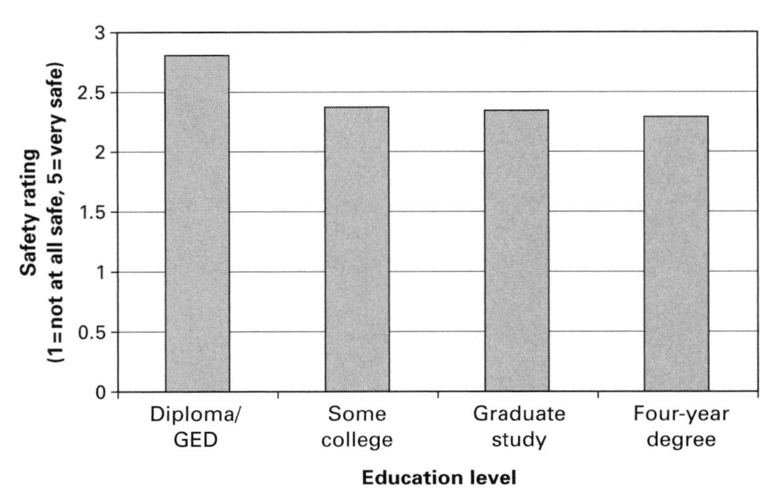

Figure 8.2
Average reported levels of safety with the identifier street address. *Note:* Diploma/ GED: mean = 2.81, sd = 1.53 ; Some college: mean = 2.37, sd = 1.34; four-year Degree: mean = 2.29, sd = 1.24; More than a four-year Degree: mean = 2.34, sd = 1.37.

- City ($F(df = 5) = 8.23$, $p < .0001$)

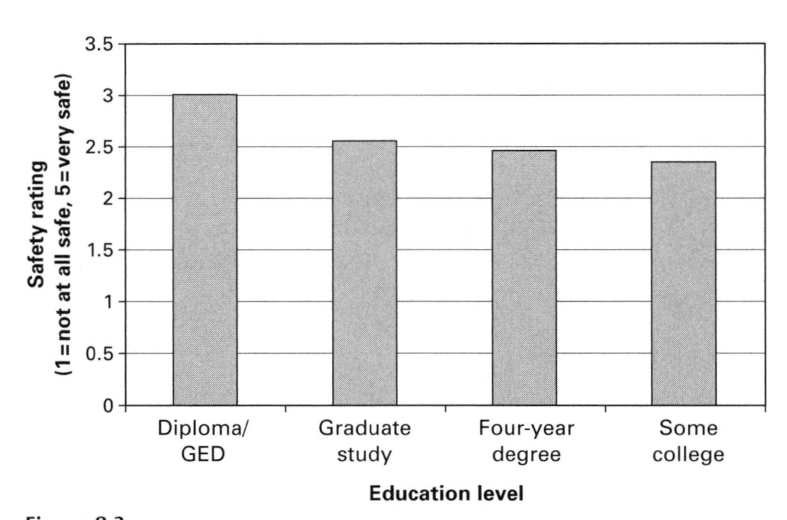

Figure 8.3
Average reported levels of safety with the identifier city. *Note:* Diploma/GED: mean = 3.00, sd = 1.54; some college: mean = 2.35, sd = 1.36; four-year degree: mean = 2.32, sd = 1.28; more than a four-year degree: mean = 2.47, sd = 1.41.

• Country (F(df = 5) = 5.58, $p < .0001$)

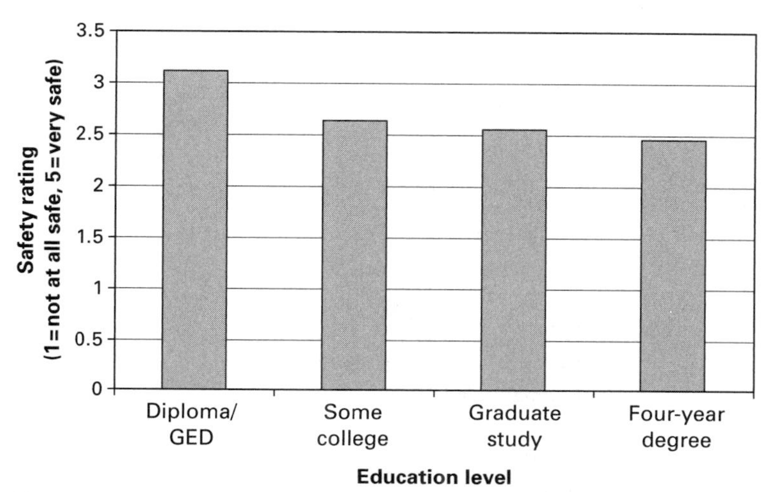

Figure 8.4

Average reported levels of safety with the identifier country. *Note:* Diploma/GED: mean = 3.11, sd = 1.55; Some college: mean = 2.64, sd = 1.56; four-year degree: mean = 2.46, sd = 1.41; more than a four-year degree: mean = 2.56, sd = 1.59.

Individuals with a high school diploma or GED reported a higher perception of safety with certain identifiers than those with a four-year college degree—for example:

• Birthplace (F(df = 5) = 4.68, $p < .0001$)

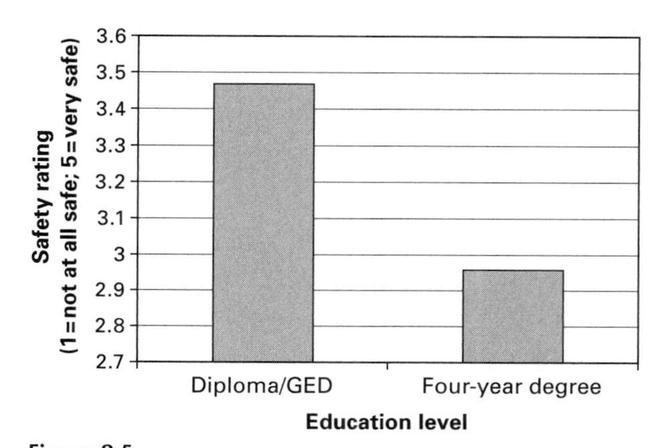

Figure 8.5

Average reported levels of safety with the identifier birthplace. *Note:* Diploma/GED: mean = 3.47, sd = 1.43; four-year degree: mean = 2.96, sd = 1.24.

- Zip code (F(df = 5) = 7.76, $p < .0001$)

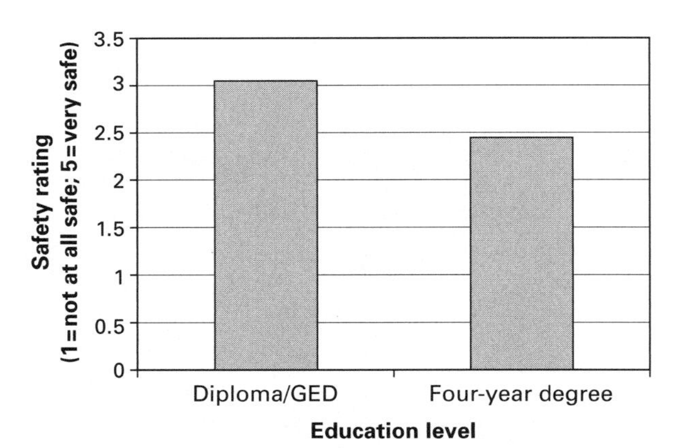

Figure 8.6
Average reported levels of safety with the identifier zip code. *Note:* Diploma/GED: mean = 3.06, sd = 1.52; four-year degree: mean = 2.44, sd = 1.22.

- Phone number (F(df = 5) = 3.40, $p = .005$)

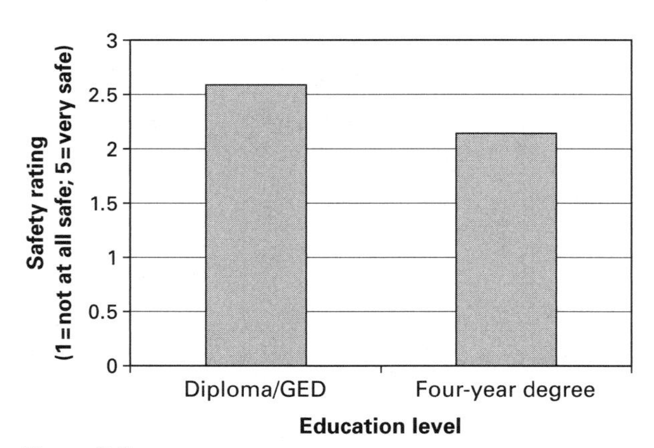

Figure 8.7.
Average reported levels of safety with the identifier phone number. *Note:* Diploma/GED: mean = 2.59, sd = 1.50; four-year Degree: mean = 2.14, sd = 1.19.

There was also a statistically significant difference between individuals with a high school diploma/GED and those with either a four-year college degree or more for:

- Social security number ($F(df = 5) = 6.07$, $p < .0001$)

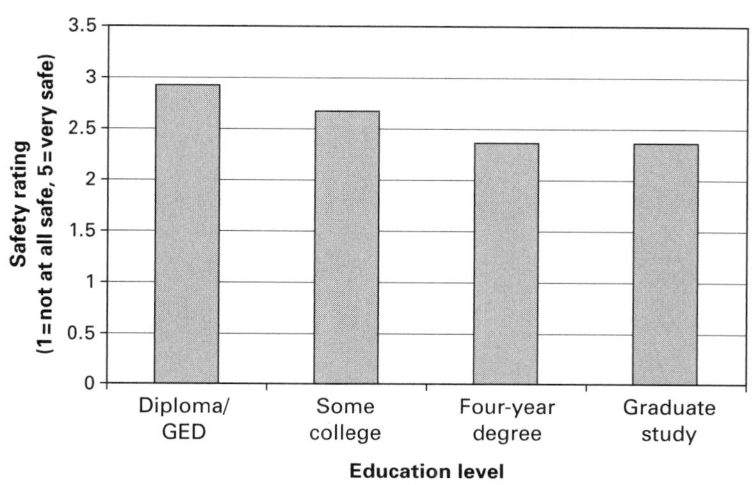

Figure 8.8.
Average reported levels of safety with the identifier social security number. *Note:* Diploma/GED: mean = 2.92, sd = 1.55; some college: mean = 2.67, sd = 1.42; four-year degree: mean = 2.36, sd = 1.29; more than a four-year degree: mean = 2.36, sd = 1.28.

- Mother's maiden name ($F(df = 5) = 4.16$, $p = .001$)

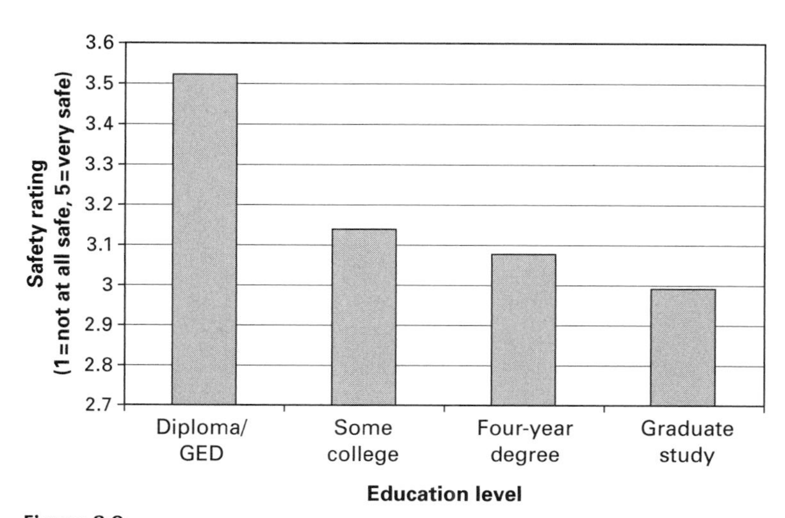

Figure 8.9.
Average reported levels of safety with the identifier mother's maiden name. *Note:* Diploma/GED: mean = 3.52, sd = 1.43; some college: mean = 3.14, sd = 1.41; four-year degree: mean = 3.08, sd = 1.275; more than a four-year degree: mean = 2.99, sd = 1.33.

• E-mail address ($F(df = 5) = 3.67$, $p = .003$)

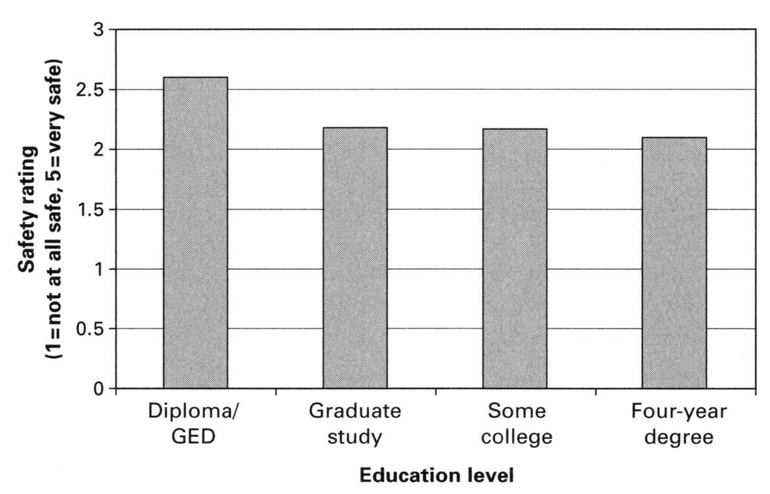

Figure 8.10

Average reported levels of safety with the identifier e-mail address. *Note:* Diploma/ GED: mean = 2.60, sd = 1.48; more than a four-year degree: mean = 2.18, sd = 1.33; some college: mean = 2.17, sd = 1.27; four-year degree: mean = 2.09, sd = 1.12.

Appendix B: Differences of Identity

Females were more concerned than their male counterparts with the possibility of:

- Unwanted telemarketing ($t_{(df = 990)}$ = −2.94, p = .003)

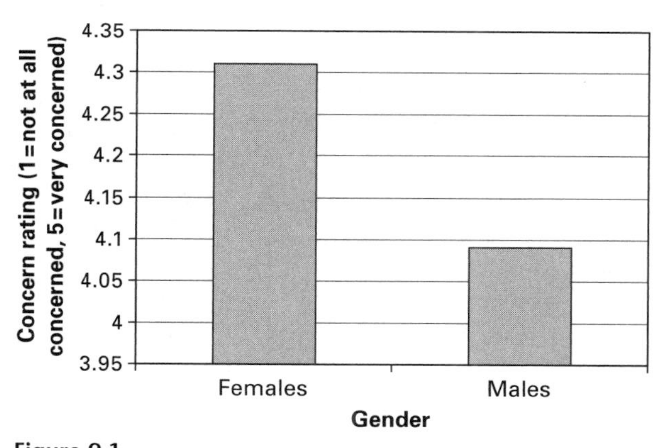

Figure 9.1
Average reported levels of concern with unwanted telemarketing. *Note:* Female: mean = 4.31, sd = 1.47; male: mean = 4.09, sd = 1.29.

- Unwanted e-mails ($t_{(df = 932)} = -3.89$, $p < .0001$)

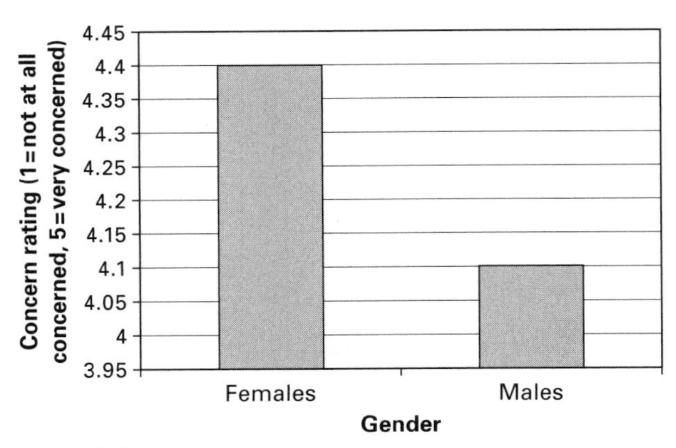

Figure 9.2

Average reported levels of concern with unwanted e-mails/spam. *Note:* Females: mean = 4.40, sd = 1.11; males: mean = 4.10, sd = 1.28.

- Another terrorist attack within the United States ($t_{(df = 993)} = -4.24$, $p < .0001$)

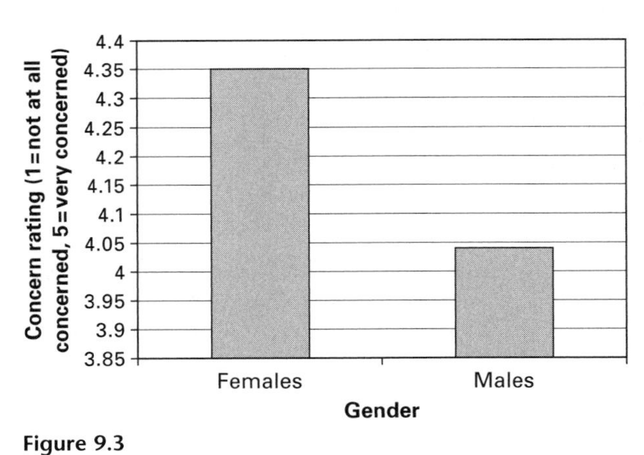

Figure 9.3

Average reported levels of concern with another terrorist attack within the United States. *Note:* Females: mean = 4.35, sd = 1.06; males: mean = 4.04, sd = 1.30.

- Identity theft ($t_{(df = 990)} = -3.01$, $p = .003$)

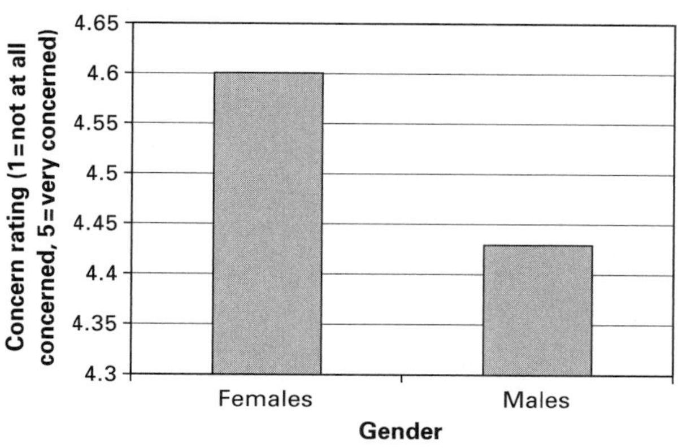

Figure 9.4
Average reported levels of concern with identity theft. *Note:* Females: mean = 4.60, sd = .84; males: mean = 4.43, sd = .93.

- The loss of civil liberties ($t_{(df = 972)} = -5.12$, $p < .0001$)

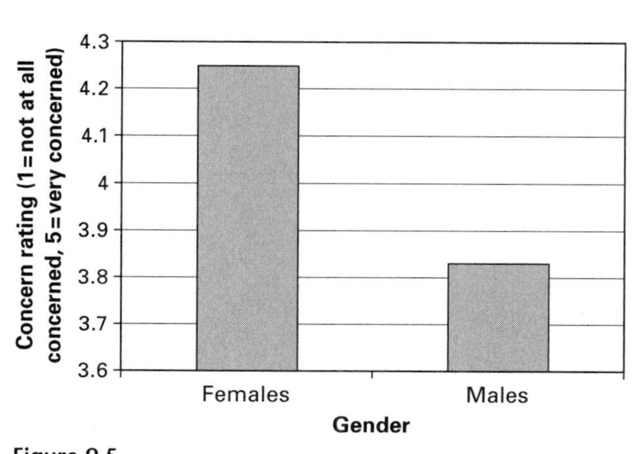

Figure 9.5
Average reported levels of concern with the loss of civil liberties. *Note:* Females: mean = 4.25, sd = 1.17; males: mean = 3.83, sd = 1.38.

Republicans were:

• More concerned than independents about another terrorist attack within the United States ($F_{(df = 4)} = 6.98$, $p < .0001$)

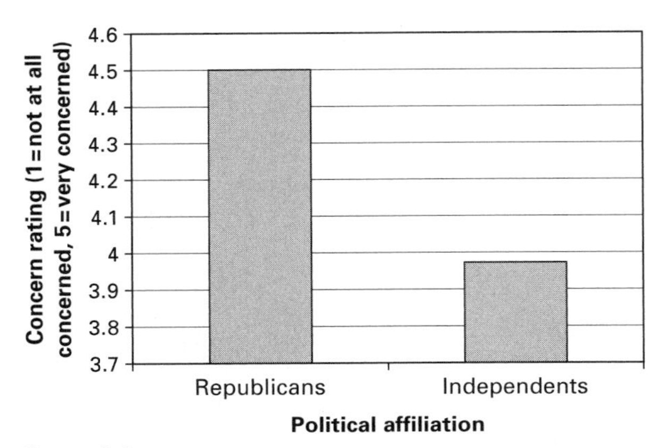

Figure 9.6
Average reported levels of concern with another terrorist attack. *Note:* Republicans: mean = 4.50, sd = .917; independents: mean = 3.97, sd = 1.28.

• More concerned than Democrats and independents about illegal aliens ($F_{(df = 4)} = 12.89$, $p < .0001$)

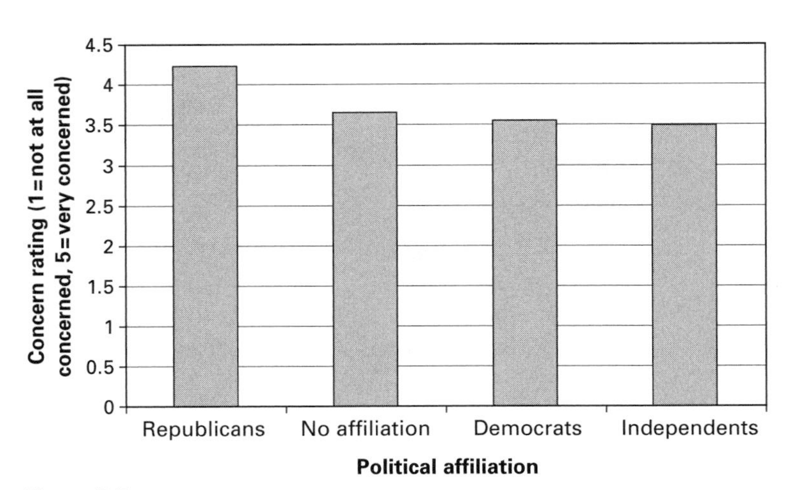

Figure 9.7
Average reported levels of concern with illegal aliens. *Note:* Republicans: mean = 4.24, sd = 1.06; no affiliation: mean = 3.65, sd = 1.39; Democrats: mean = 3.55, sd = 1.47; independents: mean = 3.50, sd = 1.37.

• Less concerned than Democrats and independents about the loss of civil liberties ($F_{(df = 4)} = 14.47$, $p < .0001$)

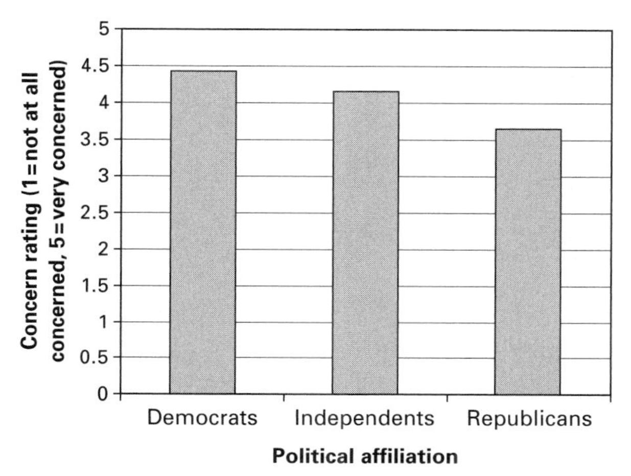

Figure 9.8

Average reported levels of concern with illegal aliens. *Note:* Democrats: mean = 4.44, sd = .10; independents: mean = 4.15, sd = 1.16; Republicans: mean = 3.65, sd = 1.44.

Individuals' level of education generated statistically significant differences in regard to their level of concern. Individuals with more than a four-year college degree were:

• Less concerned about another terrorist attack than those with a high school diploma or GED ($F_{(df = 5)} = 5.48$, $p < .0001$)

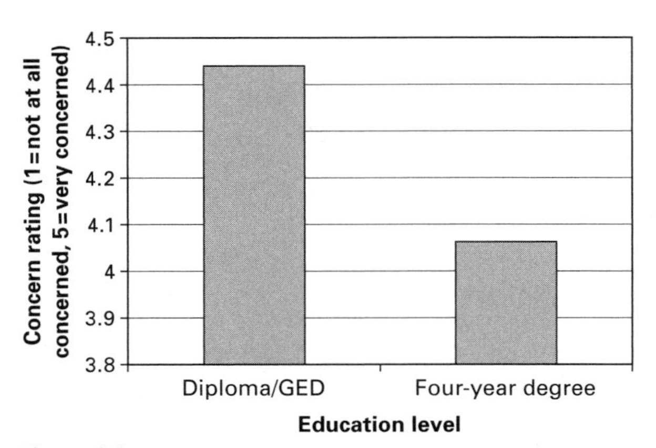

Figure 9.9

Average reported levels of concern with another terrorist attack within the United States. *Note:* Diploma/GED: mean = 4.44, sd = 1.07; four-year degree: mean = 4.06, sd = 1.24.

• Less concerned with illegal aliens than those with some college, a high school diploma or GED, and those who did not graduate from high school ($F_{(df = 5)} = 9.34$, $p < .0001$).

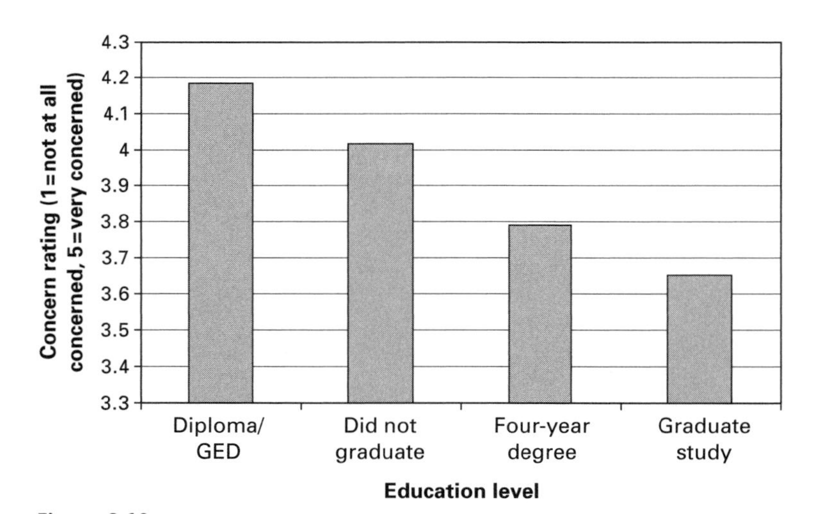

Figure 9.10

Average reported levels of concern with illegal aliens. *Note:* Diploma/GED: mean = 4.18, sd = 1.21; did not graduate: mean = 4.02, sd = 1.42; some college: mean = 3.79, sd = 1.31; four-year degree: mean = 3.65, sd = 1.37.

In addition, individuals with a four-year degree were less concerned than those with a high school diploma or GED about someone revealing personal information to the government without permission ($F_{(df = 5)} = 3.61$, $p = .003$).

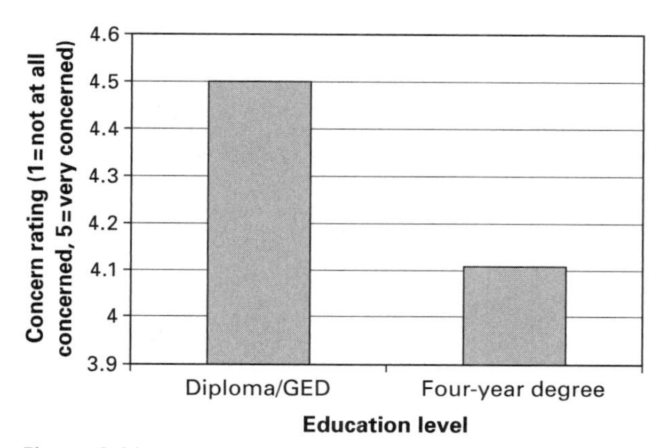

Figure 9.11
Average reported levels of concern with someone revealing personal information to the government. *Note:* Diploma/GED: mean = 4.50, sd = .10; four-year degree: mean = 4.11, sd = 1.16.

There were also statistically significant, positive correlations between age and reported levels of concern with various threats to data security. Table B.1 presents the Pearson correlation coefficients and alpha values for each variable. In addition, figure 9.1 illustrates the correlation coefficients.

Table B.1
Impact of age on levels of concern with various threats to data security ($N = 1{,}000$)

Issues	Correlation coefficients
Unwanted telemarketing	$r = .10$, $p = .002$
Unwanted e-mails or spam	$r = .11$, $p = .001$
Someone revealing your personal information to the government without your permission	$r = .10$, $p = .004$
Someone getting into your bank accounts without your permission	$r = .10$, $p = .007$
Illegal aliens in the United States	$r = .17$, $p < .0001$

Notes

Introduction

1. I am using *physiological* as it is commonly used by many in the biometrics community. Although *physiological* might more accurately refer to function, such as breathing or digestion, in relationship to biometric identifiers, it has been used to indicate functional and anatomical measurement.

2. The European Union Date Directive creates rights for persons about whom information is collected, known as "data subjects." Entities that collect information must give data subjects notice explaining who is collecting the data, who will ultimately have access to it, and why the data is being collected. Data subjects also have the right to access and correct data.

Chapter 2

1. The Alien and Sedition Acts are four different acts: Act of June 18, 1798 (Naturalization Act); Act of June 25, 1798 (Alien Friends Act); Act of July 6, 1798 (Alien Enemies Act); Act of July 14, 1798 (Sedition Act).

2. Espionage Act of 1917, 40 Stat at 219.

3. Act of Oct. 16, 1918, ch. 186, 40 Stat. 1012 (repealed June, 1952).

4. One of the journal's cartoons, Making the World Safe for Capitalism' shows a Russian absorbed in studying a paper marked "Plans for a Genuine Democracy." On one side of him Japan and England appear in a threatening attitude, and on the other Mr. Root and Mr. Russell, members of the commission sent by the United States to Russia, appear in the guise of advisers. Mr. Root has in his hands a noose, labeled "Advice," with which it is intended to entrap or choke to death the Russian Democracy. The court cannot say that the Postmaster General was not warranted in concluding that this cartoon was intended to arouse the resentment of some of our citizens of foreign birth and prevent their enlistment. In the cartoon "Congress and Big Business" Congress is represented by a disconsolate individual who is

ignored by a number of over-developed men of Big Business gathered around a table inspecting a large paper spread over it and labeled "War Plans." Congress is quoted as saying: "Excuse me, gentlemen, where do I come in?" "Big Business" replies: "Run along now; we got through with you when you declared war for us." This cartoon is intended to stir up class hatred of the war and to arouse an unwillingness to serve in the military and naval forces of the United States. The clear import is, if the war was brought on by "Big Business," then let "Big Business" carry it on, and let Labor stand aloof. The court cannot say that the Postmaster General was clearly wrong in concluding that it would interfere with enlistments. Masses Publishing Co. v. Patten, 246 F. 24 (2d Cir 1917).

5. This quote was supposed to have been published in an August 1917 issue of the *Masses,* but Postmaster General Burleson seized the issue and refused to have it mailed.

6. Masses Publishing Co. v. Patten 244 F. 535, (S.D.N.Y. 1917).

7. Masses Publishing Co. v. Patten 244 F. 535, (S.D.N.Y. 1917).

8. Masses Publishing Co. v. Patten, 246 F. 24 (2d Cir 1917).

9. Schenck v. United States, 249 U.S. 47 (1919).

10. Each of the first three counts charged the defendants with conspiring, when the United States was at war with the Imperial Government of Germany, to unlawfully utter, print, write and publish: In the first count, "disloyal, scurrilous and abusive language about the form of government of the United States;" in the second count, language "intended to bring the form of government of the United States into contempt, scorn, contumely, and disrepute;" and in the third count, language "intended to incite, provoke and encourage resistance to the United States in said war." The charge in the fourth count was that the defendants conspired "when the United States was at war with the Imperial German Government, unlawfully and willfully, by utterance, writing, printing and publication to urge, incite and advocate curtailment of production of things and products, to wit, ordnance and ammunition, necessary and essential to the prosecution of the war." The offenses were charged in the language of the act of Congress. Abrams v. United States 250 U.S. 616 (1919).

11. Ibid., 621.

12. Alien Registration Act of 1940 sec. 2.

13. Ch. 897, 54 Stat. 1201 (1940).

14. To protect financial privacy, Congress passed Gramm-Leach-Bliley Act (also known as the Financial Services Modernization Act) in 1999. Title V of the act established a set of comprehensive privacy laws at the federal level applicable to any firm that provides financial services. The law established four new requirements regarding the nonpublic personal information of a consumer. First, a financial

institution must annually disclose to consumers its policy of protection and disclosure of personal data to affiliates and nonaffiliated third parties. Second, customers can "opt-out" by preventing disclosure of personal information to third parties; there are important exceptions, such as information shared to market the financial institution's products and services. Also, "opt-out" does not apply if information sharing is necessary to a transaction or customer service. Third, a financial institution may not disclose account numbers to a third party for any marketing purposes. Finally, financial institutions must establish standards to ensure the security and confidentiality of records, protect against security threats, and protect against unauthorized access that could harm or inconvenience the subject. The act deals primarily with disclosure of information to third parties and introduces an opt-out standard.

15. http://georgewbush-whitehouse.archives.gov/nsc/infosharing/sectionV.html.

Chapter 3

1. http://www.aclu.org/technology-and-liberty/statement-barry-steinhardt-director-aclu-technology-and-liberty-program-rfid.

2. http://www.heinz.cmu.edu/~acquisti/papers/acquisti_eis_refs.pdf.

3. http://pewresearch.org/pubs/454/teens-privacy--online-social-networks.

4. Although there are distinctions drawn between various types of biometric identification systems and institutional contexts, the level of detail that a technologist or privacy advocate might expect is not presented here. The perspective that is important to understanding societal perceptions of biometric technology is not that of the expert. One might argue that the variety of biometric technologies is so diverse that discussing them in a manner less sophisticated than a technologist would obfuscate relevant distinctions. While the layperson is not as well versed in the minutiae of biometrics as a technologist, lay perceptions and opinions will continue to be important for societal acceptance. This is not to say that these generalized assessments of risks to privacy are unimportant to those developing and implementing biometric identification systems. In fact, the opposite is true: these generalizations can be used to develop technological and legal safeguards to assuage the perceived risks to privacy.

5. As Julie Cohen discusses, the theory of property has been altered by various approaches. A different possible answer, though, lies in the work of scholars who have advanced visions of property as facilitating the development of human potential. Frank Michelman, Joan Williams, and others articulate an essentially republican vision that emphasizes the role of property in ensuring an egalitarian distribution of political power and participation. In addition, Williams identifies a strain of (largely intuitive) theorizing about property that she calls the "liberal dignity"

vision. In this view, property rights may not interfere with the basic respect due all persons. Finally, Margaret Radin writes more generally of "human flourishing" in terms that encompass both individual and collective goals.Under any of these theories, property is a means to a larger end; it constitutes the individual's stake in society and undergirds society's vision of itself. These theories of property might support restrictions on the exchange of personally identified data if such restrictions are judged important to the development of community or individuality, or both.

6. Joseph Raz (1984) argues, for example, that rights-based moralities consist of rights, and those special requirements we call duties cannot account for the nature of supererogation and its role in moral life, and they do not allow personal characteristics that are virtuous or morally praiseworthy to be judged intrinsically desirable and cultivated for their own sake.

7. Privacy and security, for example, need not be at cross purposes if we liberate privacy from its individual right basis. As Peter Swire explains:

> In many instances we see "security and privacy," where the two are complementary. Under the standard approach to privacy protection, good security is an essential fair information practice. After all, good privacy policies are worth very little if hackers or other outsiders break into the system and steal the data. Both privacy and security share a complementary goal—stopping unauthorized access, use, and disclosure of personal information. Good security, furthermore, does more than keep the intruders out. It creates audit trails about which authorized users have accessed particular systems or data. These audit trails allow an accounting over time of who has seen an individual's personal information. The existence of accounting mechanisms both deters wrongdoing and makes enforcement more effective in the event of such wrongdoing. To take one example, the HIPAA medical privacy rule requires an accounting (a log) of who has seen a patient's data for other than treatment, payment, or health care operations purposes. Patients can see these logs, and the existence of the accounting mechanism will likely support both privacy (patient confidentiality) and security (prevention of unauthorized uses of the system). (Swire and Steinfeld 2002, 107)

8. As Julie Cohen (2001) has discussed, visibility is an important determinant of harm to privacy, but a persistent tendency to conceptualize privacy harms and expectations in terms of visibility has created two problems. First, focusing on visibility diminishes the salience and obscures the operation of nonvisual mechanisms designed to render individual identity, behavior, and preferences transparent to third parties. The metaphoric mapping to visibility suggests that surveillance is simply a passive observation rather than the active production of categories, narratives, and norms. Second, even a broader conception of privacy harms as a function of informational transparency is incomplete. Privacy has a spatial as well as an informational dimension. The spatial dimension of the privacy interest, which I characterize as an interest in avoiding or selectively limiting exposure, concerns the structure of experienced space. It is not negated by the fact that people in public spaces expect to be visible to others present in those spaces, and it encompasses both the arrangement of physical spaces and the design of networked communications technologies.

Chapter 4

1. According to Julie Cohen, an implicit linkage between privacy and visibility is deeply embedded in privacy doctrine. Within the common law of privacy, harms to visual privacy and harms to information privacy are subject to different requirements of proof. Of the four privacy torts, two are primarily visual and two primarily informational. The visual torts, intrusion on seclusion and unauthorized appropriation of name or likeness, require only a showing that the conduct (the intrusion or appropriation) violated generally accepted standards for appropriate behavior. The informational torts, unauthorized publication and false light, are far more stringently limited (to "embarrassing" private facts and to falsity). To make out a more general claim to information privacy, some have tried to characterize collections of personally identified data visually, likening them to "portraits" or "images," but courts have resisted the conflation of facts with faces.

2. *Whalen* v. *Roe* (1977) is a notorious decision for outlining the possible contours of a constitutional right to informational privacy. In this case, patients and doctors challenged a statute in New York that required copies of prescriptions for certain drugs to be recorded and stored in a centralized government computer, arguing that it violated their constitutional right to privacy. Although the Court rejected this claim, a majority argued that under certain circumstances, the disclosure of health care information may violate a constitutionally protected right to privacy. Justice John Paul Stevens, writing for a majority of the Court, identified informational privacy as one aspect of the constitutional right to privacy. Stevens argued, "The cases sometimes characterized as protecting 'privacy' have in fact involved at least two different kinds of interests. One is the individual interest in avoiding disclosure of personal matters, and another is the interest in independence in making certain kinds of important decisions." Privacy acts as a safeguard against government surveillance and information gathering, but only to the extent that a societal harm is prevented. In current policy debates, the use of biometric identifiers must exemplify the balance defined in Fourth Amendment jurisprudence and in the logic of *Whalen*. First, there is a recognizable interest in controlling the disclosure of information that is deemed sensitive. Second, this privacy interest is not without constraints. The interest in controlling information as a function of privacy is juxtaposed against the political objectives of protecting the common good and preventing harm. In *Whalen* the Court was persuaded that the centralized database would foster an understanding of the state's drug control laws and ultimately further the "war on drugs."

3. Consider *United States* v. *Edwards*, 498 F.2d 496, 500 (2d Cir. 1974), where the Court held: "When the risk is the jeopardy to hundreds of human lives and millions of dollars of property inherent in the pirating or blowing up of a large airplane, the danger alone meets the test of reasonableness, so long as the search is conducted in good faith for the purpose of preventing hijacking or like damage and with

reasonable scope and the passenger has been given advance notice of his liability to such a search so that he can avoid it by choosing not to travel by air."

4. This could also be understood as part of the textual meaning associated with Fourth Amendment doctrine. Although the textual reference was not explicit in the discussions of the focus group participants, the general idea defined by the doctrine of the Fourth Amendment found its way into the descriptions given socially acceptable uses of biometric technology. This point of obdurate textual meaning helps to orient the understanding given privacy. The balance between subjective expectations of privacy and the mediation of it by policy objectives is reflected in one of the longest-standing principles of privacy protection found in the jurisprudence of the Fourth Amendment.

Katz v. *United States* (1967) marked a change in the balance of privacy. In *Katz*, a bookie placed an incriminating phone call from inside a public phone booth. Law enforcement officials from the Federal Bureau of Investigation had attached an electronic listening and recording device. The question presented to the Court was whether the conversation was admissible. While both the petitioner and the government made much of the fact that the telephone from which the phone calls were placed was the determining factor in defining the boundaries of privacy, the Court saw it differently. The Court explained that whether the telephone booth was a "constitutionally protected area" was not the most important factor in determining whether a privacy violation had occurred. Rather the Court held that the "Fourth Amendment protects people, not places. What a person knowingly exposes to the public, even in his own home or office, is not a subject of Fourth Amendment protection. But what he seeks to preserve as private, even in an area accessible to the public, may be constitutionally protected." If an individual assumes that his or her conversations are private in a telephone booth, then a reasonable expectation of privacy has been formed despite the fact that the phone booth is in a place where he or she might be seen. The Court thus moved away from a material definition of privacy when contemplating the protection of the Fourth Amendment against search and seizure and argued that "the electronic device employed to achieve that end did not happen to penetrate the wall of the booth can have no constitutional significance." The concurrence points to the potential problem of the *Katz* analysis and the movement of the doctrine because of flux in social perceptions. Justice Harlan argued that if the Fourth Amendment protects people, not places, an unresolved question is what type of protection is afforded in what types of situations. He wrote, "Generally, as here, the answer to that question requires reference to a 'place.'" Harlan anticipated the potential problem of freeing privacy from its physical moorings. He wrote, "Conversations in the open would not be protected against being overheard, for the expectation of privacy under the circumstances would be unreasonable." As a solution, Harlan articulated what has become the prevailing standard in deciding Fourth Amendment cases. In essence, the two-pronged test requires the presence of a subjective expectation of privacy that society deems objectively reasonable. The Supreme Court has applied this principle to hold that a Fourth Amendment search does not occur unless "the individual manifested a subjective expectation of privacy in the object of the challenged search," and "society [is] willing to recognize that expectation as reasonable." The decision lends insight into the nexus of subjective and objective expectations of privacy and the balancing that occurs. Because Fourth Amendment doctrine applies to governmental searches and seizures, there is a balancing test, which attempts to retain individual privacy while facilitating a governmental objective.

5. Observation is used here not to imply physical observance but rather to describe the possible gathering of biometric data to track physical activities, transactions, or interactions.

Chapter 5

1. Trust and confidence and their decline in the modern era have been the topic of much discussion in academia, media, and the government. This might seem to be an ironic point of debate because since the constitutional founding, distrust of government has been central to political and economic liberalism as a limiting concept. The development of liberal thought was built on a healthy distrust of government. For example, David Hume (1985), an influential political theorist of the 1700s and an important intellectual resource for the constitutional debate, remarked that "political writers have established it as a maxim, that, in contriving any system of government, and fixing the several checks and controls of the constitution, every man ought to be supposed a knave, and to have no other end, in all his actions, than private interest" (42). Well-known political theorists such as Hume, James Madison, and John Locke distrusted the frailties of political institutions when populated with men who would invariably be tempted by the aphrodisiac of power. Distrust of the *publicus* was a common concern among the Enlightenment philosophers, whose writings informed the thinking of our founding fathers. As Hobbes (1994) argued, "A plain husbandman is more prudent in affairs of his own house than a privy councilor in the affairs of another man" (40).

2. The Privacy Rule establishes, for the first time, a foundation of federal protections for the privacy of protected health information. The rule does not replace federal, state, or other law that grants individuals even greater privacy protections, and covered entities are free to retain or adopt more protective policies or practices.

3. The GLB Act gives authority to eight federal agencies and the states to administer and enforce the Financial Privacy Rule and the Safeguards Rule. These two regulations apply to financial institutions, which include not only banks, securities firms, and insurance companies but also companies that provide many other types of financial products and services to consumers.

4. http://www.prospect.org/cs/articles?article=bowling_together.

5. As 2001 ended, Americans were more united, readier for collective sacrifice, and more attuned to public purpose than we have been for several decades. Indeed, we have a more capacious sense of "we" than we have had in the adult experience of most Americans now alive. The images of shared suffering that followed the terrorist attacks suggested a powerful idea of cross-class, cross-ethnic solidarity. Americans also confronted a clear foreign enemy, an experience that both drew us closer to one another and provided an obvious rationale for public action. In the aftermath of September's tragedy, a window of opportunity has opened for a sort of civic renewal that occurs only once or twice a century (Putnam 2002).

6. Putnam (2002) wrote "We were especially surprised and pleased to find evidence of enhanced trust across ethnic and other social divisions. Whites trust blacks more, Asians trust Latinos more, and so on, than these very same people did a year ago. An identical pattern appears in response to classic questions measuring social distance: Americans in the fall of 2001 expressed greater open-mindedness toward intermarriage across ethnic and racial lines, even within their own families, than they did a year earlier. To be sure, trust toward Arab Americans is now about 10 percent below the level expressed toward other ethnic minorities. We had not had the foresight to ask about trust in Arab Americans a year ago, so we cannot be certain that it has declined, but it seems likely that it has. Similarly, we find that Americans are somewhat more hostile to immigrant rights. Other surveys have shown that public skepticism about immigration increased during 2001, but that trend may reflect the recession as much as it does the terrorist attacks. Yet despite signs of public support for antiterrorist law-enforcement techniques that may intrude on civil liberties, our survey found that Americans are in some respects more tolerant of cultural diversity now than they were a year ago. Opposition to the exclusion of 'unpopular' books from public libraries actually rose from 64 percent to 71 percent. In short—with the important but partial and delimited exception of attitudes toward immigrants and Arab Americans—our results suggest that Americans feel both more united and more comfortable with the nation's diversity. We also found that Americans have become somewhat more generous, though the changes in this domain are more limited than anecdotal reports have suggested. More people in 2001 than in 2000 reported working on a community project or donating money or blood. Occasional volunteering is up slightly, but regular volunteering (at least twice a month) remains unchanged at one in every seven Americans. Compared with figures from immediately after the tragedy, our data suggest that much of the measurable increase in generosity spent itself within a few weeks."

Chapter 6

1. Consider the host of paternalistic acts that Vandeveer (1986), one of the prominent thinkers on paternalism, says are part and parcel of our public policy landscape:

1. Legal requirements that motorcyclists wear helmets
2. Legal requirements prohibiting self-medication
3. Legal prohibitions on the possession or use of certain risk-laden drugs
4. Laws requiring the testing of drugs by the Food and Drug Administration prior to their legalization in the United States
5. Prohibitions on how long a minor may work or access certain types of work
6. Curfews on those below a certain age
7. Prohibitions on gambling
8. Compulsory education of the young
9. Restrictions by brokerage firms on which adults may engage in purchasing stocks or commodities
10. The indoctrination of young children with political or religious doctrines
11. Attempts to disallow prisoners from volunteering for biomedical experiments
12. Attempts to disallow minors and mentally disabled persons from being subjects of biomedical experiments or organ donations
13. Attempts to prevent persons sentenced to capital punishment from undergoing it even when they demand it
14. Legal prohibition on suicide
15. Legal prohibitions on voluntary euthanasia
16. Psychiatrist pressuring patients into sexual relations to "uncover areas of sexual blocking"
17. Deception of patients by physicians, or other health care personnel, to avoid disturbing the patient or to facilitate their consent to a procedure the physician believes is desirable for the patient
18. Use of drugs on patients to make them more compliant and receptive to a procedure the physician believes to be in the patient's best interest
19. Legally required vaccinations
20. Compulsory participation in systems providing for adequate income on retirement
21. Required courses at a university
22. Prohibitions on voluntary self-enslavement
23. Civil commitment of those judged dangerous to themselves
24. Involuntary sterilization
25. Distribution of welfare in kind rather than in cash
26. Involuntary blood transfusions of those opposed to them on either moral or religious grounds.
27. Deception of children "for their benefit," for example, telling them that a medical procedure only hurts a little when that is false or perpetuating the Santa Claus myth.
28. Fluoridation of water supplies
29. Prohibitions on purchase of explosives and poisons
30. Required waiting periods for divorce
31. Medical disqualifications of student athletes from playing sports
32. Infanticide of radically defective infants
33. Force-feeding of "hunger strikes"
34. Punishment to rehabilitate the punished
35. Labor laws restricting minimum wages, maximum hours, or who may work.
36. The arrest of drunks to prevent their being "rolled"

2. The calculus is strikingly simple according to Mill:

1. S's liberty to do X is infringed only if S desires to do X.
2. S does not desire to do X.
3. S's liberty to do X is not infringed.

As Vandeveer (1968) observes, Mill's calculus for evaluating paternalistic actions is valid but presents a potential problem: "Being at liberty to perform an act is more appropriately characterized in terms of the absence of constraints of barriers to certain possible courses of action – independently of whether I do or will desire to perform them" (30). Vandeveer has redefined paternalism in this manner:

1. A deliberately does (or omits) X
2. A believes that his (her) doing (or omitting) X is contrary to S's operative preference, intention or disposition at the time A does (or omits) X [or when X affects S-or would have affected S if X had been done (or omitted) X]
3. A does (or omits) X with the primary or sole aim of promoting a benefit for S [a benefit that, A believes, would not accrue to S in the absence of A's doing (or omitting) X] or preventing a harm to S in the absence of A's doing (or omitting) X]. (Vandeveer, 1968, 30)

3. Of course, the risks of identity theft and terrorism are not entirely separable. In the 9/11 Commission Report (2004), for example, impersonation is cited as a key tool for terrorists. The report emphasizes that fraud is no longer just a problem of theft: "At many entry points to vulnerable facilities, including gates for boarding aircraft, sources of identification are the last opportunity to ensure that people are who they say they are."

4. Joel Feinberg (1983) describes the calculation of risk as involving a series of considerations: (1) the degree of probability that harm to oneself will result from a given course of action; (2) the seriousness of the harm being risked, that is, the value or importance of that which is exposed to the risk; (3) the degree of probability that the goal inclining one to shoulder the risk will in fact result from the course of action; (4) the value or importance of achieving that goal, that is, just how worthwhile it is to one (this is the intimately personal factor, requiring a decision about one's own preferences, that makes the reasonableness of risk-assessment on the whole so difficult for the outsider to make); and (5) the necessity of risk, that is the availability or absence of alternative, less risky, means to the desired goal.

5. Policy protections are not the way in which societal acceptance of biometric systems of identification can be fostered. Technological innovations can also enhance the security of biometric systems against vulnerabilities. Innovations in biometric system security include measures that protect the vulnerabilities at each point of interaction with a biometric system of identification. In the case of facial recognition, for example, do eyeglasses or facial hair affect the efficiency of the system? Quality control can also be a point of vulnerability when a good image is classified as poor and a poor-quality image is classified as good. Template creation and storage also introduces the problem of function creep or identity theft, when a template created for one purpose is then used for another purpose unintended by the individual.

Some technological innovations that are designed to combat these vulnerabilities and strengthen the protections include watermarking; encoded encrypted techniques that embed a secret code into a biometric template that can be decrypted only with the image of the enrolled individual; and revocable biometrics. (Jain and Uludag 2003). For example, revocable biometrics is a purposeful distortion of the biometric image with known properties during enrollment. When the individual represents his or her biometric, it must be distorted in exactly the same way as it was during enrollment. Watermarking is another technological safeguard that hides an individual biometric in a variety of images in order to protect it. Encryption can be also be used to protect biometric templates. During authentication, the encrypted templates are decrypted with the correct key.

Chapter 7

1. HIPAA was passed in 1996 to facilitate data transfer of health information. Subsequently, the Department of Health and Human Services (HHS) developed federal privacy standards to protect individuals' medical records and health information provided to medical professionals and health care providers. In accordance with previous U.S. privacy policies, the standards provide individuals with access to their medical records, in addition to control over access to and use of that information. The HHS privacy rule limits the circumstances under which identifying information can be disclosed to third parties without explicit consent from the individual. Barring these specified circumstances, written authorization from the data subject must be obtained prior to disclosure of health information, and no conditions of treatment or benefits may be attached to receipt of this authorization. Under the HHS standards, patients generally have the right of access to their medical information on request from medical professionals and health care systems and must opt in through authorized consent before information is shared. The type of consent protected in HIPAA involves an individual's written authorization for any use or disclosure of patient information that is not for treatment, payment, or health care operations, or is otherwise permitted or required, is required under HIPAA. These authorizations must be plainly stated and contain information regarding the disclosure to be made. In addition to these requirements, the persons disclosing and receiving the information, the duration of the permission, and the right to revoke in writing are also outlined. HIPAA also limits the information gathered to "the minimum necessary to accomplish the intended purpose, use, disclosure or request." This statutory language creates an obligation for entities to ensure accountability and transparency in the handling of personal information. To ensure compliance, a covered entity must develop and implement policies and procedures to reasonably limit uses and disclosures to the minimum necessary. To implement these restrictions on uses and disclosures, these policies and procedures must include details on the restrictions to access and uses of information, must identify the persons or classes of persons who may access the information, the categories of personal information, and any conditions under which they may have access to this information. Under the provisions of HIPAA, policies and procedures must also be put in place for recurring disclosures and requests for disclosures. HIPAA meets this obligation

with the requirement of a written acknowledgment from patients of receipt of the privacy practices notice.

To protect financial privacy, Congress passed the Gramm-Leach-Bliley Act (also known as the Financial Services Modernization Act) in 1999. Title V of the act established a set of comprehensive privacy laws at the federal level applicable to any firm that provides financial services. The new law established four new requirements regarding the nonpublic personal information of a consumer. First, a financial institution must annually disclose to consumers its policy of protection and disclosure of personal data to affiliates and nonaffiliated third parties. Second, customers can opt out by preventing disclosure of personal information to third parties; there are important exceptions, such as information shared to market the financial institution's products and services. Also, "opt out" does not apply if information sharing is necessary to a transaction or customer service. Third, a financial institution may not disclose account numbers to a third party for any marketing purposes. Finally, financial institutions must establish standards to ensure the security and confidentiality of records, protect against security threats, and protect against unauthorized access that could harm or inconvenience the subject. The GLBA deals primarily with disclosure of information to third parties and introduces an opt-out standard. Under GLB, a disclosure of "nonpublic personal information" cannot be made to another corporation without providing the consumer with a notice. The privacy policy must give the policies for sharing data with both affiliates and nonaffiliated third parties, including the categories of information that may be disclosed.

On this point, many have criticized the notice requirement of GLB to be burdensome for financial institutions and ineffective for providing privacy protection for consumers. This criticism arises because the form of consent associated with GLB is an opt-out provision. With receipt of the privacy notice, the consumer has the opportunity to opt out of having his or her information shared with nonaffiliates under GLB. Privacy advocates have argued that this is an ineffective way to protect consumer privacy and, more important, is at odds with a consent provision because often the consumer is unaware of which he or she is opting out of. Industry has generally favored a default rule of allowing sharing, while privacy advocates have generally favored a default rule prohibiting sharing unless the individual specifically opts in. HIPAA and GLB also address the necessary provisions for security. GLB provides, in part, that "it is the policy of the Congress that each financial institution has an affirmative and continuing obligation to respect the privacy of its customers and to protect the security and confidentiality of those customers' nonpublic personal information." Administrative, technical, and physical safeguards are mandated under GLB to protect the security and confidentiality of customer records and information.

2. *OECD Guidelines on the Protection of Privacy and Transborder Flows of Personal Data*, Organization for Economic Co-operation and Development (Sept. 23, 1980), http://www.oecd.org/documentprint/0,2744,en_2649_34255_1815186_1_1_1_1,00.html.

3. Federal Trade Commission, *Privacy Online: A Report to Congress* (June 1998), *at* http://www.ftc.gov/reports/privacy3/toc.htm.

4. Asia-Pacific Economic Cooperation, *APEC Privacy Principles: Consultation Draft* (March 31, 2004), at http://www.export.gov/apeccommerce/privacy/consultation-draft.pdf.

5. *Id.*

6. European Parliament Directive 95/46/EC 1995 O.J. (L 281).

7. *Id.* art. 6(1)(b).

8. *Id.* art. 6(1)(e).

9. *Id.* art. 17(1).

10. *Id.* art. 8(1).

11. Joel R. Reidenberg and Paul M. Schwartz. "Data Protection Law and On-Line Services: Regulatory Responses." 8.

12. *Id.* art. 12(c).

13. *Id.* arts. 14, 22.

References

Abrams v. United States. 1919. 250 U.S. 616.

Acquisti, A. 2004a. Security of personal information and privacy: Technological solutions and economic incentives. In *The economics of information security*, ed. J. Camp and R. Lewis, 165–178. Norwell, MA: Kluwer.

Acquisti, A. 2004b. *The economics of privacy*. http://www.heinz.cmu.edu/~acquisti/economics-privacy.html (accessed May 5, 2008).

Agre, P. E. 2001. *Technology and privacy: The new landscape*. Cambridge, MA: MIT Press.

Akrich, M. 1992. A summary of a convenient vocabulary for the semiotics of human and nonhuman assemblies. In *Shaping technology—building society: Studies in sociotechnical change*, ed. W. Bijker and J. Law, 259–624. Cambridge, MA: MIT Press.

American Civil Liberties Union. 1938. *Thumbs down! The fingerprinting menace to civil liberties*. New York: American Civil Liberties Union.

American Civil Liberties Union. 2001, September 19. Letter to the Senate Commerce Committee on S. 2949, the Aviation Security Improvement Act. http://www.aclu.org/privacy/gen/15120leg20020919.html (accessed May 5, 2008).

American Civil Liberties Union. 2004. ACLU testifies to Congress on dangers of biometric passports. *Privacy International*. http://www.privacyinternational.org/article.shtml?cmd%5B347%5D=x-347-60594.

Associated Press. 2004, December 26. *Sign on Sandiego*. http://www.signonsandiego.com/uniontrib/20041226/news_1n26lapd.html (accessed May 20, 2008).

Austin, J. 1995/1832. *The province of jurisprudence determined*. Ed. W. Rumble. Cambridge: Cambridge University Press.

Ayres, I., and J. Braithwaite. 1992. *Responsive regulation*. New York: Oxford University Press.

Beardsley, E. 1971. Privacy: Autonomy and selective discourse. In *NOMOS XIII*, ed. R. P. Chapman, 169–181. New York: Atherton.

Benn, S. 1984. Privacy, freedom, and respect for persons. In *Philosophical Dimensions of Privacy*, ed. F. Schoeman, 223–245. Cambridge: Cambridge University Press.

Bertillon, A. 1896. *Signalment instructions including the theory and practice of anthropometrical description*. Kila, MT: Kessinger.

Big Blind Brother. 2003, September 14. *Tampa abandons facial recognition policing*. http://www.cfif.org/htdocs/freedomline/current/in_our_opinion/facial_recognition.html (accessed June 16, 2009).

Bijker, W., and J. Law. 1992a. Strategies, resources, and the shaping of technology. In *Shaping technology/building society*, ed. W. E. Bijker and J. Law, 103–107. Cambridge, MA: MIT Press.

Bijker, W., and J. Law. 1992b. What catastrophe tells us about technology and society. In *Shaping technology/building society*, ed. W. Bijker and J. Law, 1–19. Cambridge, MA: MIT Press.

Bijker, W., and T. Pinch. 1987. The social construction of facts and artifacts: Or how the sociology of science and the sociology of technology might benefit each other. In *The social construction of technological systems*, ed. W. Bijker. Cambridge, MA: MIT Press.

Biometrics.gov. 2007, January. Hand geometry. http://www.biometrics.gov/ (accessed May 22, 2008).

Bourdieu, P. 1991. *Language and symbolic power*. Cambridge, MA: Harvard University Press.

Bourdieu, P. 1992. *An invitation to reflexive sociology*. Chicago: University of Chicago Press.

Bowker, G. C., and S. L. Star. 1999. *Sorting things out: Classification and its consequences*. Cambridge, MA: MIT Press.

Bowman, L. 2003, August 21. Tampa drops face recognition system. *CNET news*. http://www.news.com/Tampa-drops-face-recognition-system/2100-1029_3-5066795.html (accessed May 20, 2008).

Brandeis, B., and L. Warren. 1890. The right to privacy. *Harvard Law Review* 4(1): 193–220.

Brownsword, R. 2008. *Rights, regulation, and technological revolution*. New York: Oxford University Press.

Buchanan, A. E. 1983. Medical paternalism. In *Paternalism*, 61–82. Minneapolis: University of Minnesota Press.

Buck v. Bell. 1927. 274 U.S. 2007.

Caplan, J. 2001. This or that particular person: Protocols of identification in nineteenth-century Europe. In *Documenting individual identity: The development of state practices in the modern world,* ed. J. Caplan and J. Torpey. Princeton, NJ: Princeton University Press, 49–67.

Carlson, W. B. 2001. The telephone as political instrument: Gardiner Hubbard and the formation of the middle class in America, 1875–1880. In *Technologies of power,* ed. M. T. Allen and G. Hecht, 1–24. Cambridge, MA: MIT Press.

Castells, M. 2000. *The rise of the network society.* Oxford: Blackwell.

CBS. 2005, October 2. Privacy rights under attack. *CBS News.* Shttp://www.cbsnews.com/stories/2005/09/30/opinion/polls/main894733.shtml (accessed June 30, 2007).

cbsnews. 2005.Privacy rights under attack. cbsnews.com/stories (accessed 2005).

Chafee, Z. 1941. *Free speech in the United States.* Cambridge, MA: Harvard University Press.

Cheng, L. 1984. Free, indentured, enslaved: Chinese prostitutes in nineteenth century America. In *Labor immigration under capitalism: Asian workers in the United States before World War I,* ed. L. Cheng and E. Bonacich, 402–434. Berkeley: University of California Press.

Clarke, R. 1988. Information technology and dataveillance. *Communications of the ACM* 31:498–512.

Cohen, J. 1998. *Lochner in cyberspace: The new economics orthodoxy of "rights management." Michigan Law Review* 97:462–563.

Cohen, J. 2000. Examined lives: Informational privacy and the subject as object. *Stanford Law Journal* 52:1373–1437.

Cole, S. 2001. *Suspect identities: A history of fingerprinting and criminal identification.* Cambridge, MA: Harvard University Press.

Coleman, J. S. 1988. Social capital in the creation of human capital. *American Journal of Sociology* 94:95–120.

Coleman, J. S. 1990. *Foundations of social theory.* Cambridge, MA: Belknap Press of Harvard University Press.

Collier, P. 2001. Biometric advocacy report 111/17. *Biometrics Advocacy Report* 3(7). http://ibia.vwh.net/newslett04201.htm.

Cook, K., and R. Hardin. 2001. Norms of cooperativeness and networks of trust. In *Social norms,* ed. M. Hechter and K.-D. Opp. New York: Russell Sage Foundation.

Cook, K., R. Hardin, and M. Levi. 2005. *Cooperation without trust*. New York: Russell Sage Foundation.

Curiel, J. 2006, June 25. *The last days of privacy*. http://articles.sfgate.com/2006-06-25/opinion/17299190_1_fingerprint-system-credit-cards-computer-system.

Das, T., and B.-S. Teng. 1998. Between trust and confidence: Developing trust in partner cooperation in alliances. *Academy of Management Review* 23:491–512.

Dennis v. United States. 1951. 341 U.S. 494.

De Tocqueville, A. 1954. *Democracy in America*, ed. Rev. F. Bowen and P. Bradley. New York: Vintage Books.

Dewey, J. 1954. *Public and its problems*. Athens, OH: Swallow Press.

Durkheim, E. 1893/1964. *Division of labour in society*. New York: Free Press.

Dworkin, R. 1978. *Taking rights seriously*. Cambridge, MA: Harvard University Press.

Dworkin, R. 1985. Law's ambitions for itself. *Virginia Law Review* 71:173–181.

Eagleton, T. 1991. *Ideology: An introduction*. London: Verso.

Electronic Frontier Foundation. Who's watching you? http://www.eff.org/wp/biometrics-whos-watching-you (accessed March 31, 2010).

Electronic Frontier Foundation. 2008a. Biometric identifiers. http://epic.org/privacy/biometrics/ (accessed June 27, 2008).

Electronic Frontier Foundation. 2008b. Stop the spying. *NSA Spying*. http://www.eff.org/issues/nsa-spying (accessed June 18, 2008).

Ericson, R., and K. Haggerty. 1997. *Policing the risk society*. Oxford: Clarendon Press.

Etzioni, A. 1999. *The limits of privacy*. New York: Basic Books.

Faden and Beauchamp. 1986. *A history and theory of informed consent*. New York: Oxford University Press.

Feder, B. 2001,. Technology and media: A surge in demand to use biometrics. *New York Times,* December 17.

Feinberg, J. 2003. *Problems at the roots of law: Essays in legal and political theory*. New York: Oxford University Press.

Ferdinand, D. 1992. *Privacy and social freedom*. Cambridge: Cambridge University Press.

Fischer, C. 1992. *America calling: A social history of the telephone*. Berkeley: University of California Press.

Flanagan, M., D. Howe, and H. Nissenbaum. 2008. Embodying values in technology. In *Information technology and moral philosophy,* ed. J. Van De Hoeven and J. Weckert, 322–353. Cambridge: Cambridge University Press.

Foucault, M. 1986. Of other spaces (1967), Heterotopias. *Diacritics* 16:22–27.

Fraser, N., and L. Gordon. 1994. A genealogy of "dependency": Tracing a keyword of the U.S. welfare state. *Signs. Journal of Women in Culture and Society* 19:309–336.

Fried, A. 1997. *McCarthyism: The great American red scare.* New York: Oxford University Press.

Friedman, J. S. 2005. *The secret histories: Hidden truths that challenged the past and changed the world.* New York: Macmillan.

Fukuyama, F. 1995. *Trust: Social virtues and the creation of prosperity.* New York: Free Press.

Fukuyama, F. 1996. *Trust: Social virtues and the creation of prosperity.* New York: Free Press.

Gadamer, H. G. 1979. *Truth and method.* London: Sheed and Ward.

Galton, F. 1908. *Memories of my life.* London: Methuen.

Gavison, R. 1984. *Philosophical dimensions of privacy.* Cambridge: Cambridge University Press.

Gibbs, M. R. 2002. Information privacy and data quality: The problem of database fragmentation. In *Proceedings of the Third Australian Institute of Computer Ethics Conference,* Sydney, 30 September.

Giddens, A. 1991. *Modernity and self-identity.* Oxford: Polity Press.

Gilliom, J. 2001. *Overseers of the poor: Surveillance, resistance, and the limits of privacy.* Chicago: University of Chicago Press.

Gordon, W. 1993. A property rights in self-expression: Equality and individualism in the natural law of intellectual property. *Yale Law Journal* 102:1533–1609.

Green, N., and S. Smith. 2003. "A spy in your pocket"? The regulation of mobile data in the UK. *Surveillance and Mobilities* 1(4). http://www.surveillance-and-society.org/journalv1i4.htm.

Griffin, R. J. 2003. *The faces of anonymity: Anonymous and pseudonymous publication from the sixteenth to the twentieth century.* New York: Palgrave.

Griswold v. Connecticut. 1965. 381 U.S. 479.

Habermas, J. 1989. *Structural transformation of the public sphere.* Cambridge, MA: MIT Press.

Halperin, D. 2005. *Social capital.* Oxford: Polity Press.

Hardin, R. 1999. Democracy and collective bads. In *Democracy's edges,* ed. I. Shapiro and C. Hacker-Cordon. Cambridge: Cambridge University Press, 63–83.

Hardin, R. 2006. *Trust.* Cambridge: Polity Press.

Hart, H. L. A. 1963. *Law, liberty and morality.* Stanford: Stanford University Press.

Hart, H. L. A. 1983. *Jhering's heaven of concepts and essays in jurisprudence and philosophy.* Oxford: Clarendon Press, 1983.

Hart, H. L. A.1984. *Essays in jurisprudence and philosophy.* New York: Oxford University Press.

Hecht, G. 2001. Technology, politics, and national identity in France. In *Technologies of power,* ed. M. Allen and G. Hecht, 253–293. Cambridge, MA: MIT Press.

Herreros, F. 2004. *The problem of forming social capital: Why trust?* Princeton, NJ: Princeton University Press.

Herschel, W. 1916. *The origin of fingerprinting.* New York: Oxford University Press.

Hobbes, T. 1957. *Leviathan.* Oxford: Basil Blackwell.

Hobbes, T. 1994. *The Clarendon Edition of the works of Thomas Hobbes.* Ed. N. Malcolm. New York: Oxford University Press.

Homer. 1963. *The odyssey.* New York: Doubleday.

Hoyt, E. 1969. *The Palmer raids: 1919–1920: An attempt to suppress dissent.* New York: Seabury Press.

Hume, D. 1985. *Essays: Moral, political and literary.* Ed. E. F. Miller. Indianapolis: Liberty Classics.

IC International. 2002. *Public attitudes toward the uses of biometric identification technologies by government and the private sector.* London: IC International.

International Biometrics Group. 2002. *The biometrics industry: One year after 9/11.* September. http://www.ibgweb.com/9-11.html (accessed June 17, 2008).

Jain, A. 2004. *The Proceedings of International Conference on Pattern Recognition.* Cambridge, UK, August 23–26.

Jain, A. K., P. Flynn, and A. A. Ross. 2008. *The handbook of biometrics.* New York: Springer.

Jain, A., A. Ross, and S. Prabhakar. 2004. An introduction to biometric recognition. *IEEE Transactions on Circuits and Systems for Video Technology* 14(1): 4–20.

James, S. H. 2005. *Forensic science: An introduction to scientific and investigative techniques*. Boca Raton, FL: CRC Press.

Jasanoff, S. 2004. Afterword. In *States of knowledge: The co-production of science and social order*, ed. S. Jasanoff, 274–282. London: Routledge.

Jeffreys-Jones, R. 2007. *The FBI: A history*. New Haven, CT: Yale University Press.

Jupiter Research. N.d. Online privacy. http://www.jmm.com/ (accessed July 11, 2008).

Kaluszynski, M. 2001. Republican identity: Bertillonage as government technique. In *Documenting individual identity: The development of state practices in the modern world*, ed. J. Caplan and J. Torpey. Princeton, NJ: Princeton University Press, 123–138.

Katz v. United States. 1967. 389 U.S. 347.

Kern, S. 1983. *The culture of time and space, 1880–1918*. Cambridge, MA: Harvard University Press.

Kline, R. 2005. Resisting consumer technology in rural America: The telephone and electrification. In *How users matter: The co-construction of users and technology*, ed. N. Oudshoorn and T. Pinch, 51–66. Cambridge, MA: MIT Press.

Kramer, R., and K. Cook. 2004. Trust and distrust in organizations: Dilemmas and approaches. In *Trust and distrust in organizations: Dilemmas and approaches*, ed. R. Kramer and K. S. Cook, 1–18. New York: Russell Sage Foundation.

Latour, B. 1988. *The pasteurization of France*. Cambridge, MA: Harvard University Press.

Latour, B. 2005. *Reassembling the social: An introduction to actor-network-theory*. New York: Oxford University Press.

Lessig, L. 2000. *Code and other laws of cyberspace*. New York: Basic Books.

Levi, M., and L. Stoker. 2000. Political trust and trustworthiness. *Annual Review of Political Science* 3:457–507.

Lianos, M. 2003. *Social control after Foucault*. http://www.surveillance-and-society.org/articles1(3)/AfterFoucault.pdf (accessed February 15, 2008).

Litan, P. S. 1998. *None of your business: World data flows, electronic commerce and the European data directive*. Washington, DC: Brookings Institute.

Little, M. 2003. Hammering their way through the barriers. In *Training the excluded for work: Access and equity for women, immigrants, First Nations, youth, and people with low income*, ed. M. Cohen, 108–123. Vancouver: University of British Columbia Press.

Locke, J. 1952. *Second treatise on government.* Upper Saddle River, NJ: Prentice Hall.

Loury, G. 1977. A Dynamic theory of racial income differences. In *Women, minorities and employment discrimination,* ed. P. Wallace. Lanham, MD: Lexington Books, 153–186.

Lyon, D. 2001. Under my skin: From identification papers to body surveillance. In *Documenting individual identity: The development of state practices in the modern world,* ed. J. Caplan and J. Torpey. Princeton, NJ: Princeton University Press, 291–310.

Lyon, D. 2007. *Surveillance studies: An overview.* Oxford: Polity Press.

MacDonald, A. 1902. *A plan for the study of man.* Washington, DC: U.S. Government Printing Office.

Marx, G. 1990. *Undercover: Police surveillance in America.* Berkeley: University of California Press.

Masses Publishing Company v. Patton. 1917. 244 Fed. 535 (S.D.N.Y.).

Marx, G. T. 2001. Identity and anonymity. In *Documenting individual identity: The development of state practices in the modern world,* ed. J. Caplan and J. Torpey. Princeton, NJ: Princeton University Press, 311–327.

McCahill, M. 1998. Beyond Foucault: Towards a contemporary theory of surveillance. In *Surveillance, closed circuit television and social control,* ed. C. Norris, J. Moran, and G. Armstrong. Aldershot: Ashgate, 41–65.

McIntyre v. Ohio Election Commission. 1995. 514 U.S. 334.

Medema, N. M. 1997. *Economics and the law: From Posner to post-modernism.* Princeton, NJ: Princeton University Press.

Mercuro, N., and S. G. Medema. 1997. *Economics and the law: From Posner to post-modernism.* Princeton, NJ: Princeton University Press.

Messmer, E. 2007. *Palm print biometrics aid Atlanta police.* November 15. http://www.pcworld.com/article/id,139668-c,biometricsecuritydevices/article.html (accessed May 23, 2008).

Michael, L. 2008. *Biometric security for mobile banking.* Washington, DC: World Resources Institute.

Mill, J. S. 1869. *On liberty.* London: Longman.

Mintie, D. 2007. Biometric watch. February 28. http://www.biometricwatch.com/BW_in_print/mad_cow_biometrics.htm (accessed May 20, 2008).

Most, M. N.d. The identity crisis: What comes after the post-9/11 hype? http://www.idworldonline.com/ (accessed June 27, 2008).

Munzer, S. 1990. *A theory of property*. Cambridge: Cambridge University Press.

NAACP v. Alabama ex rel. Patterson. 1958. 357 U.S. 449.

National Commission on Terrorist Attacks upon the United States. 2004. *The 9/11 Commission Report: Final report of the National Commission on Terrorist Attacks upon the United States*. New York: Norton.

Noiriel, G. 2001. The identification of citizen: The birth of republican civil status in France. In *Documenting individual identity: The development of state practices in the modern world*, ed. J. Caplan and J. Torpey. Princeton, NJ: Princeton University Press, 28–49.

North, M. 2003. *Anonymous renaissance*. Chicago: University of Chicago Press.

Olmstead v. United States. 1928. 277 U.S. 438.

O'Reilly, K. 1989. *Racial matters: The FBI's secret file on Black America, 1960–1972*. New York: Free Press.

O'Sullivan, O. 1997. Biometrics come to life. http://www.banking.com/aba/cover_0197.htm (accessed May 20, 2008).

Oudshoorn, N., and T. Pinch. 2005a. How users and non-users matter. In *How users matter: The co-construction of users and technology*, ed. N. Oudshoorn and T. Pinch, 2–25. Cambridge, MA: MIT Press.

Oudshoorn, N., and T. Pinch. 2005b. How users matter: Introduction. In *How users matter: The co-construction of users and technology*, ed. N. Oudshoorn and T. Pinch. Cambridge, MA: MIT Press.

Parenti, C. 2004. *The soft cage: Surveillance in America from slavery to the War on Terror*. New York: Basic Books.

Pew Research. 2007. Teens, privacy and on-line social networks. http://pewresearch.org/pubs/454/teens-privacy-online-social-networks (accessed March 19, 2008).

Phillips, P., P. Rauss, and S. Der. 1996. *FacE REcognition Technology (FERET) recognition algorithm development and test report*. Adelphi, MD: U.S. Army Research Laboratory.

Posner, R. 1984. An economic theory of privacy. In *Philosophical dimensions of privacy: An anthology*, ed. F. Schoeman, 333–345. Cambridge: Cambridge University Press.

Prosser, W. 1984. Privacy: A legal analysis. In *Philosophical dimensions of privacy*, ed. F. Schoeman, 104–154. Cambridge: Cambridge University Press.

Putnam, R. 1993. *Making democracy work*. Princeton, NJ: Princeton University Press.

Putnam, R. 2000. *Bowling alone*. New York: Simon & Schuster.

Putnam, R. 2002. *Democracies in flux: The evolution of social capital in contemporary society*. New York: Oxford University Press.

Raz, J. 1984. Rights based moralities. In *Theories of rights*, ed. J. Waldron, 182–200. New York: Oxford University Press.

Regan, P. 1995. *Legislating privacy: Technology, social values and public policy*. Chapel Hill: University of North Carolina Press.

Rhodes, K. 2004. Aviation security: Challenges in using biometric technology. Testimony before the Subcommittee on Aviation, Committee on Transportation and Infrastructure, House of Representatives. May 19.

Robinson, G. 2001. *By order of the president: FDR and the internment of Japanese Americans*. Cambridge, MA: Harvard University Press.

Roiter, J. 2008, April 21. Keystroke recognition aids online authentication at credit union. http://searchfinancialsecurity.techtarget.com/news/article/0,289142,sid185_gci1310519,00.html (accessed May 20, 2008).

Rosen, J. 2005. *The naked crowd*. New York: Random House.

Rule, J. B. 2007. *Privacy in peril*. Oxford: Oxford University Press.

Samuelson, P. 2000. Privacy as intellectual property. *Stanford Law Review* 52:1125–1173.

Sander, T., and R. Putnam. 2002. Walking the civic talk after September 11. *Christian Science Monitor*, February 19. http://www.csmonitor.com/2002/0219/p11s02-coop.html (accessed May 5, 2008).

Scanlon, T. M. 1984. Rights, goals, and fairness. In *Theories of rights*, ed. J. Waldron. New York: Oxford University Press, 136–152.

Scanlon, T. M. 2003. *The difficulty of tolerance: Essays in political philosophy*. Cambridge, MA: Harvard University Press.

Schenck v. United States. 1919. 249 U.S. 47.

Schmidt, R. 2000. *Red scare: FBI and the origins of anticommunism in the United States 1919–1943*. Copenhagen: Museum Tusculanum Press.

Schoeman, F. 1984. Privacy: Philosophical dimensions of the literature. In *Philosophical dimensions of privacy: An anthology*, ed. F. Schoeman, 1–32. Cambridge: Cambridge University Press.

Schoeman, F. D. 1992. *Privacy and social freedom*. Cambridge: Cambridge University Press.

Schrecker, E. 2002. *The age of McCarthyism: A brief history with documents*. New York: St. Martin's Press.

Shklar, J. 1990. *The faces of injustice.* New Haven, CT: Yale University Press.

Silver, I. 1979. *The crime control establishment.* Englewood Cliffs, NJ: Prentice Hall.

Siegrist, M., T. Earle, and H. Gutscher. 2007. Trust in cooperative risk management. In *Trust in cooperative risk management,* ed. M. Siegrist, T. Earle, and H. Gutscher, 1–49. London: Earthscan.

Sismondo, S. 2008. Science and technology studies and an engaged program. In *The handbook of science and technology studies,* ed. E. Hackett, O. Amsterdamska, and M. Lynch, 14–26. Cambridge, MA: MIT Press.

Slobogin, C. 2007. *Privacy at risk: The new government surveillance.* Chicago: Chicago University Press.

Smith, J. M. 1956. *Freedom's fetters: The Alien and Sedition Laws and American Civil liberties.* Ithaca, NY: Cornell University Press.

Smith, P. 1984. *The rise of industrial America.* New York: McGraw-Hill.

Solove, D. 2004. *Digital person.* New York: New York University Press.

Solove, D. 2008. Do social networks bring the end of privacy? *Scientific American,* http://www.scientificamerican.com/article.cfm?id=do-social-networks-bring (accessed March 31, 2008).

Solove, D. J. 2009. *Understanding privacy.* Cambridge, MA: Harvard University Press

Solzhenitsyn, A. 1991. *The cancer ward.* New York: Farrar, Straus and Giroux.

Spikerman, S., J. Grossklags, and B. Berendt. 2000. EPrivacy in 2nd generation e-commerce. In *Third Conference on Electronic Commerce,* 38–47. Norwell, MA: Kluwer.

Steinhardt, B. 2004. Director of the ACLU Technology and Liberty Program, on RFID tags before the Commerce, Trade and Consumer Protection Subcommittee of the House Committee on Energy and Commerce. http://www.aclu.org/technology-and-liberty/statement-barry-steinhardt-director-aclu-technology-and-liberty-program-rfid (accessed March 19, 2008).

Stigler, G. 1950. The development of utility theory. *Journal of Political Economy* 58:307–321.

Stone, G. 2004. *Perilous times: Free speech in wartime from the Sedition Act of 1798 to the War on Terrorism.* New York: Norton.

Swire, P. 2009. The Obama administration's silence on privacy. *New York Times,* June 2. http://bits.blogs.nytimes.com/2009/06/02/the-obama-adminstrations-silence-on-privacy/#more-10523 (accessed June 25, 2009).

Swire, P., and L. Steinfeld. 2002. Security and privacy after September 11: The health care example. *Minnesota Law Review* 86:101–126.

Taylor, S., and P. Todd. 1995. Assessing IT usage: The role of prior experience. *MIS Quarterly* 19:561–570.

Temple-Raston, D. 2009. Biometrics play new role in passport technology, March 19. http://www.npr.org/templates/story/story.php?storyId=102056426 (accessed June 24, 2009).

Theoharis, A. 2004. *The FBI and American democracy: A brief critical history*. Lawrence: University Press of Kansas.

Thomas v. Collins. 1945. 323 U.S. 516.

Tyler, T. 2006. *Why people obey the law*. Princeton, NJ: Princeton University Press.

Tyler, T. R., and P. DeGoey. 1996. Trust in organizational authorities: The influence of motive attributions on willingness to accept decisions. In *Trust in organizations: Frontiers of theory and research,* ed. R. Kramer and T. Tyler, 331–356. Thousand Oaks, CA: Sage.

United States v. Edwards. 1974. 415 U.S. 800.

United States v. Rumely. 1953. 345 U.S. 41.

Van de Veer, D. 1986. *Paternalistic intervention: The moral bounds of benevolence*. Princeton, NJ: Princeton University Press.

Venkatesh, F., and D. Davis. 2000. A theoretical extension of the technology acceptance model: Four longitudinal field studies. *Management Science* 46:186–204.

Visionics. 2001. A surge in the demand to use biometrics. http://www.biometricgroup.com/in_the_news/ny_times.html (accessed March 18, 2009).

Waldron, J. 1984. Introduction. In *Theories of Rights*, ed. J. Waldron, 1–20. New York: Oxford University Press.

Walzer, M. 1984. Liberalism and the art of separation. *Political Theory* 12: 315–330.

Walzer, M. 2003. A better vision: The idea of civil society. In *The civil society reader,* ed. V. Hodgkinson and M. Foley, 306–321. Hanover, NH: University Press of New England.

Watner, C., and W. McElroy. 2004. *National identification systems: Essays in opposition*. Jefferson, NC: McFarland.

Wayman, A. J. 2005. *Biometric systems: Design, technology and performance*. New York: Springer.

Wayman, J., A. Jain, D. Maltoni, and D. Maio. 2005. *An introduction to biometric authentication systems*. New York: Springer.

Weinrub, L. 2000. The right to privacy. *Social Philosophy and Policy* 17(2): 25–44.

Weinstein, M. 1971. The uses of privacy in the good life. In *NOMOS XIII*, ed. R. P. Chapman, 99–113. New York: Atherton Press.

Westin, A. 1970. *Privacy and freedom*. N.p.: Bodley Head Ltd.

Whalen v. Roe. 1977. 429 U.S. 589.

White, J. A. 2000. *Democracy, justice, and the welfare state: Reconstructing public care*. University Park: Pennsylvania State University Press.

White House. N.d. *NSIS sharing information with the private sector*. http://www.whitehouse.gov/nsc/infosharing/sectionV.html (accessed June 18, 2008).

Wilder, H., and B. Wentworth. 1918. *Personal identification*. Boston: R. G. Badger.

Wittgenstein, L. 1968. *Philosophical investigations*. New York: Macmillan.

Wittgenstein, L. 1978. *Philosophical grammar*. Berkeley: University of California Press.

Woodward, J. 2001. *Facing up to biometric technology*. Santa Monica, CA: RAND.

Woodward, J. N. M. 2003. *Identity assurance in the information age*. New York: McGraw-Hill.

Woodward, J., N. Orlans, and P. Higgens. 2003. *Identity assurance in the information age*. New York: McGraw-Hill.

Yack, B. 1992. *The longing for total revolution*. Berkeley: University of California Press.

Yar, M. 2003. Panoptic power and the pathologisation of vision: Critical reflections on the Foucauldian thesis. *Surveillance and Society* 1(3): 254–271.

Zhang, D. 2004. *Palmprint identification*. Norwell, MA: Kluwer.

Zuriek, E., and K. Hindle. 2004. Governance, security and technology: The case of biometrics. *Studies in Political Economy* 73:113–137.

Index